Professional English (Materials)

专业英语（材料类）

何丽红　主编

西南交通大学出版社
·成　都·

图书在版编目（CIP）数据

专业英语：材料类 = Professional English (Materials) / 何丽红主编. —成都：西南交通大学出版社，2019.8（2022.7 重印）
ISBN 978-7-5643-7121-0

Ⅰ. ①专… Ⅱ. ①何… Ⅲ. ①材料科学 – 英语 – 高等学校 – 教材 Ⅳ. ①TB3

中国版本图书馆 CIP 数据核字（2019）第 188765 号

Professional English (Materials)
专业英语（材料类）
Zhuanye Yingyu (Cailiaolei)

何丽红 / 主　编

责任编辑 / 牛　君
封面设计 / 何东琳设计工作室

西南交通大学出版社出版发行
（四川省成都市金牛区二环路北一段 111 号西南交通大学创新大厦 21 楼　610031）
发行部电话：028-87600564　　028-87600533
网址：http://www.xnjdcbs.com
印刷：四川森林印务有限责任公司

成品尺寸　185 mm × 260 mm
印张　9.75　　字数　297 千
版次　2019 年 8 月第 1 版　　印次　2022 年 7 月第 2 次

书号　ISBN 978-7-5643-7121-0
定价　28.00 元

课件咨询电话：028-87600533
图书如有印装质量问题　本社负责退换

前言 || PREFACE

“专业英语”是材料类专业学生在学完公共英语和专业基础课程后可选修的一门课程，主要目的是使学生通过学习，能掌握材料类专业常用的英语词汇，能较流畅地阅读、理解和翻译有关的科技英语文献和资料，并掌握英文论文的书写格式及英文论文摘要的写作技巧，从而使学生进一步提高英语应用能力，并为后期学习、科研和工作有意识地利用所学知识，通过阅读最新的专业英语文献，能跟踪学科的发展动态，为从事创新性的工作打下基础。

本书分两部分，第一部分是基础篇，根据材料总体介绍和材料类别（金属材料、陶瓷、聚合物和复合材料）分成五个单元，每个单元根据内容不同含若干课程，每个课程又由一篇课文和一篇阅读材料组成，共30篇，主要介绍材料类基础专业知识，使学生掌握相关专业英文词汇，锻炼阅读专业文献的能力；第二部分是提高篇，从学生掌握科技英语论文为出发点，分“查、读、写、投”四个单元，使学生对科技英语论文检索、阅读、撰写以及投稿有全面的认识与掌握。此外，在本书最后附录部分将材料专业常用元素、物质英文名称等进行了整理与总结，方便初学者对材料专业英语词汇的掌握。

本书内容较多，虽然作为校内自编教材已采用十一载，经过多次反复修订，但限于编者水平，书中仍可能出现错漏，还望读者不吝指正。

编 者

2019年4月

目录 || CONTENTS

PART I

PART Ⅱ

PART I

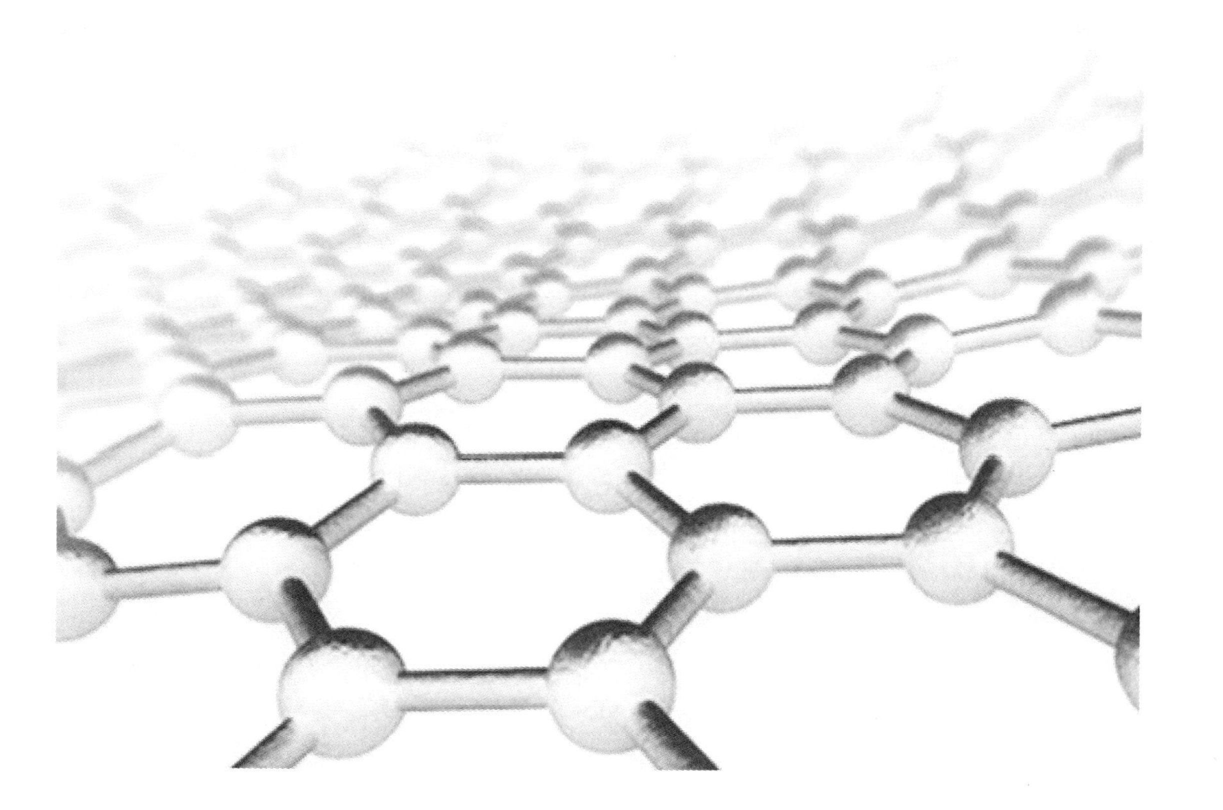

UNIT 1 INTRODUCTION TO MATERIALS

Lesson 1 Materials Science and Engineering

Life in the 21st century is ever dependent on an unlimited variety of ***advanced materials***. In our consumptive world, it is easy to take for granted the ***macro-***, ***micro-***, and ***nanoscopic*** building blocks that comprise any item produced. Materials are properly more deep-seated in our culture than most of us realize. Transportation, housing, clothing, communication, recreation and food production—virtually every segment of our everyday lives is influenced on one degree or another by materials. Historically, the development and advancement of societies have been intimately tied to the members' abilities to produce and manipulate materials to fill their needs. In fact, early civilizations have been designated by the level of their materials development (i.e. Stone Age, ***Bronze Age***, Iron Age). Today, materials play a decisive role in all technological changes.

The earliest humans has access to only a very limited number of materials, those occur naturally stone, wood, clay, skins, and so on. With time they discovered techniques for producing materials that had properties superior to those of the natural ones; these new materials included ***pottery*** and various metals. Furthermore, it was discovered that the properties of a material could be altered by heat treatments and by the addition of other substances. At this point, materials utilization was totally a selection process, that is, deciding from a given, rather limited set of materials the one that was best suited for an application by virtues of its characteristic. It was not until relatively recent times that scientists came to understand the relationships between the structural elements of materials and their properties. This knowledge, acquired in the past 60 years or so, has empowered them to fashion, to a large degree, the characteristics of materials. Thus, tens of thousands of different materials have evolved with rather specialized characteristics that meet the needs of our modern and complex society.

Materials

The term material may be broadly defined as any solid-state component or device that may be used to address a current or future societal need. For instance, simple building materials such as nails, wood, coatings, etc. address our need of shelter. Other more intangible materials such as nanodevices may not yet be widely proven for particular applications, but will be essential for the future needs of our civilization. Although the above definition includes solid nanostructural building blocks that assemble to form larger materials, it excludes complex liquid compounds such as crude oil, which may be more properly considered a precursor for materials.

Materials Science and Engineering

Materials science is an ***interdisciplinary*** study that combines chemistry, physics, ***metallurgy***, engineering and very recently life sciences. One aspect of materials science involves studying and designing materials to make them useful and reliable in the service of humankind. It strives for basic understanding of how structures and processes on the atomic scale result in the properties and functions familiar at the engineering level. Materials scientists are interested in physical and chemical phenomena acting across large magnitudes of space and time scales. In this regard it differs from physics or chemistry where the emphasis is more on explaining the properties of pure substances. In materials science there is also an emphasis on developing and using knowledge to understand how the properties of materials can be controllably designed by varying the compositions, structures, and the way in which the bulk and surfaces phase materials are processed.

In contrast, materials engineering is, on the basis of those structure properties correlations, designing or engineering the structure of a material to produce a predetermined set of properties. In other words, materials engineering mainly deals with the use of materials in design and how materials are manufactured.

As schematized in Figure1.1, there are four main aspects materials science and technology: synthesis, manufacturing and processing, composition and structure, properties and performances. The behavior in manufacture and in use coupled with economic factors characterizes the performance of a material. Closely linked are four aspects of materials science. The material is elaborated during synthesis (polymer) or manufacturing (metals, alloys, ***ceramics***, etc.). Processing concerns the shaping of a material and the preparation of a finished object according to its behavior. For example, the production of a car body involves successively rolling of the sheet steel from a bar of steel, the stamping of the sheet steel to form the body and a series of finishing operations (painting, etc.).

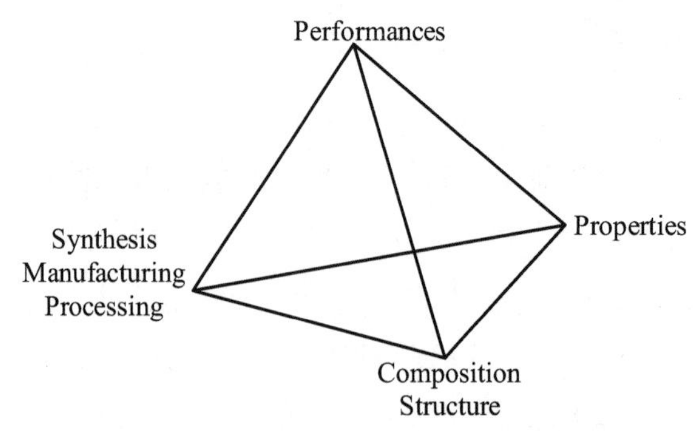

Figure 1.1 The four basic aspects of materials science and technology

To obtain optimal properties, it is essential to master the structure and composition of the materials and consequently to have access to a series of the ***sophisticated analysis techniques***.

It is the numerous contributions of Materials Science and Technology, which has completely remodeled the world, which supports us by freeing man of a huge number of constraints, linked to our environment. Our way of life has been radically transformed within a few decades largely due to the contributions of Materials Science and Engineering which lead to the creation of the tools of the modern life: automobiles, aircraft, bridge, cable cars, computers, ***telecommunications*** equipment, satellites etc.

New Words and Expressions

advanced materials [əd'vɑːnst mə'tɪərɪəlz] 先进材料

Bronze Age [brɔnz eidʒ] 铜器时代

macro- ['mækrəʊ] 宏观的，大的

micro- ['maɪkrəʊ] 微观的，小的

nanoscopic 纳米级，纳米观

pottery ['pɒtəri] *n.* 陶器，陶器制造（术）

interdisciplinary [ˌɪntə'dɪsəplɪnəri] *adj.* 交叉学科的，跨学科的

metallurgy [mə'tælədʒi] *n.* 冶金学

ceramics [sɪ'ræmɪks] *n.* 陶瓷，陶瓷制品

sophisticated analysis technique 精密分析技术

telecommunication [ˌtelɪkəˌmjuːnɪ'keɪʃn] *n.* 电信，长途通信，无线电通信

Reading Material

Classification of Materials

Materials are classified according to various criteria such as their composition, their structure or their properties. Here, distinction is made between three large groups of materials (Figure 1.2). This classification is based on the atomic structures and the nature of bonds:

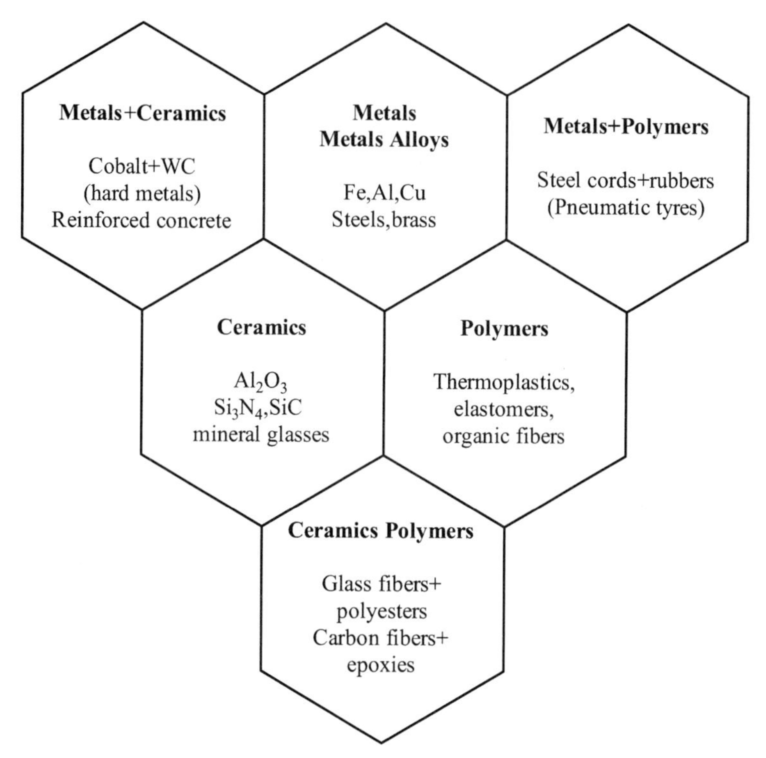

Figure 1.2 The three classes of materials: metals, ceramics and polymers With some possible combinations of composite materials

Metals and their alloys (metallic bonding);

Ceramics (ionic bonding and covalent bonding);

Polymers (covalent bonding and secondary bonding).

This classification can be examined with the help of ***the Periodic Table of the Elements***.

The majority of the elements are metals (approximately 70%) (to the left and in the centre of Mendeleev's table). The non-metals, such as oxygen occupy the right hand side of the Periodic Table. In the intermediate region between metals and non-metals, there occurs a certain number of elements such as carbon and silicon (***semiconductor***) which escape this simple classification.

Metals: Metals (including alloys) consist of atoms and are characterized by metallic bonding (i.e., the valence electrons of each atom are delocalized and shared among all the atoms). Most of the elements in the Periodic Table are metals. Examples of alloys are Cu-Zn (brass), Fe-C (steel), and Sn-Pb (solder). Alloys are classified according to the majority element present. The main classes of alloys are iron-based alloys for structures; copper-based alloys for piping, utensils, thermal conduction, electrical conduction, etc.; and aluminum-based alloys for lightweight structures and metal-matrix composites. Alloys are almost always in the ***polycrystalline*** form.

Ceramics: Ceramics are inorganic compounds such as Al_2O_3(for spark plugs and for substrates for microelectronics), SiO_2(for electrical insulation in microelectronics), Fe_3O_4(ferrite for magnetic memories used in computers), silicates (clay, cement, glass, etc.), and SiC (an ***abrasive***). The main classes of ceramics are oxides, carbides, nitrides, and silicates. Ceramics are typically partly crystalline and partly ***amorphous***. They consist of ions (often atoms as well) and are characterized by ionic bonding and often covalent bonding.

Polymers: Polymers in the form of thermoplastics (nylon, polyethylene, polyvinyl chloride, rubber, etc.) consist of molecules that have covalent bonding within each molecule and ***van der Waals' forces*** between them. Polymers in the form of ***thermosets*** (e.g., epoxy, ***phenolics***, etc.) consist of a network of covalent bonds. Polymers are amorphous, except for a minority of thermoplastics. Due to the bonding, polymers are typically electrical and thermal insulators. However, conducting polymers can be obtained by doping, and conducting polymer-matrix composites can be obtained by the use of conducting fillers.

Composites: Composite materials are ***multiphase*** materials obtained by artificial combination of different materials to attain properties that the individual components cannot attain. An example is a lightweight structural composite obtained by embedding continuous carbon fibers in one or more orientations in a polymer matrix.

Any classification of materials possesses such an arbitrary character, for there is no discontinuity between the three classes of materials. Other classifications, based on specific material properties such as semi-conductivity, can be justified.

Materials that are utilized in high-technology applications are sometimes termed advanced materials. By high technology we mean a device or product that operates or functions using relatively intricate and sophisticated principles; examples include electronic equipment (VCRs, CD players, etc.), computers, ***fiberoptic*** systems, spacecraft, air-craft, and military rocketry. These advanced materials are typically either traditional materials whose properties have been enhanced or newly developed, high-performance materials. Furthermore, they may be of all material types (e.g. metals, ceramics, polymers), and are normally relatively expensive.

New Words and Expressions

cobalt ['kəubɔːlt] *n.* 钴（符号为 Co）
reinforced concrete [ˌriːɪn'fɔːst, -'fəurst] *n.* 钢筋混凝土
brass [brɑːs] *n.* 黄铜
pneumatic tyre [nuː'mætɪk 'taiə] *n.* 气胎，内车胎
mineral ['mɪnərəl] *n.* 矿物，矿石；*adj.* 矿物的，似矿物的
thermoplastic [ˌθɜːməu'plæstɪk] *adj.* 热塑性的；*n.* 热塑性塑料
elastomer [ɪ'læstəmə(r)] *n.* 弹性体，人造橡胶
polyester [ˌpɒli'estə(r)] *n.* 聚酯
epoxy [ɪ'pɒksi] *adj.* 环氧的；*n.* 环氧树脂
the Periodic Table of the Elements 元素周期表
semiconductor [ˌsemikən'dʌktə(r)] *n.* 半导体
polycrystalline [pɒlɪ'krɪstəlaɪn] *adj.* 多晶的
abrasive [ə'breɪsɪv] *n.* 研磨剂
amorphous [ə'mɔːfəs] *adj.* 无定形的，无组织的，非结晶的
van der Waals' forces 范德华力
thermoset ['θɜːməset] *adj.* 热固性的；*n.* 热固性塑料
phenolics [fɪ'nɒlɪks] *n.* 酚醛塑料
multiphase ['mʌltɪfeɪz] *n.* 多相
fiberoptic ['fɪbəɒptɪk] *n.* 光纤

Lesson 2 Structural Characteristic of Materials

The arrangement of atoms in solids, in general, in particular, will exhibit ***long-range order***, only ***short-range order***, or a combination of both. Solids that exhibit long-range order are referred to as crystalline solids, while those in which that periodicity is lacking are known as amorphous, glassy, or ***noncrystalline*** solids.

The difference between the two is best illustrated schematically, as shown in Figure

1.3. From the figure it is obvious that a solid possesses long-range order when the atoms repeat with a periodicity that is much greater than the bond lengths. Most metals and ceramics, with the exception of glasses and glass-ceramics, are crystalline.

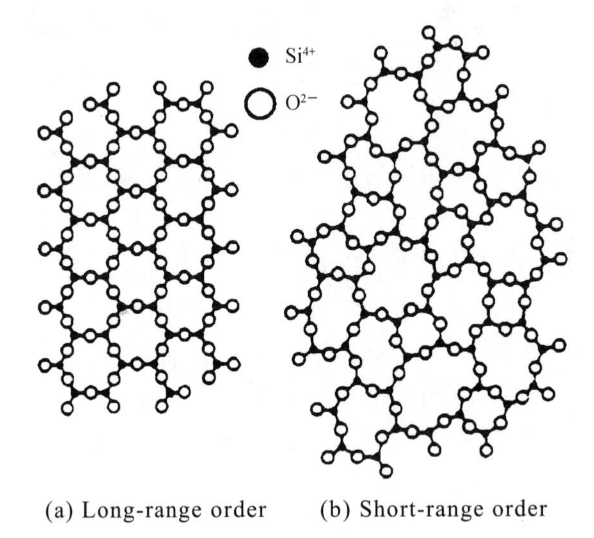

(a) Long-range order (b) Short-range order

Figure 1.3 The difference between long-range order and short-range order

Crystalline bodies remain solid, i.e. retain their shape, up to a definite temperature (***melting point***) at which they change from the solid to liquid state. During cooling, the inverse process of ***solidification*** takes place. In both cases, the temperature remains constant until the material is completely melted or respectively solidified.

Amorphous bodies, when heated, are gradually softened in a wide temperature range and become viscous and only then change to the liquid state. In cooling, the process takes place in the opposite direction.

The crystalline state of a solid is more stable than amorphous state.

Examples of such changes from amorphous to crystalline state are the turbidity effect appearing in ***inorganic*** glasses on heating or in optical glasses after a long use, partial crystallization of molten amber on heating, or additional crystallization and strengthening of ***nylon fibres*** on tension.

Crystalline bodies are characterized by an ordered arrangement of their elementary particles (ions, atoms or molecules). The properties of crystals depend on the electronic structure of atoms and the nature of their interactions in the crystal, on the ***spatial arrangement*** of elementary particles, and on the composition, size and shape of crystals.

The structure of crystals is described by using the concepts of fine structure and micro-and macro-structure depending on the size of structural components and the methods employed to reveal them.

Microscopic examinations make it possible to determine the size and shape of grains (crystals), the presence of crystals of different nature, their distribution and relative volume quantities, the shape of foreign inclusions and ***microvoids***, ***orientations*** of crystals, and some special ***crystallographic*** characteristics (twins, slip lines, etc.).

Macrostructure of crystals is studied by the naked eye or with a magnifying glass. This method can reveal the pattern of a fracture, shrinkage cavities and voids, and the shape and size of large crystals. It is also possible to detect cracks, chemical ***inhomogeneities***, fibrous textures, etc. by using specially prepared (polished and etched) specimens. Macrostructure examination is a valuable method for studying crystalline materials.

Crystal Structures

As noted above, long-range order requires that atoms be arrayed in a three-dimensional pattern that repeats. The simplest way to describe a pattern is to describe a ***unit cell*** within that pattern. A unit cell is defined as the smallest region in space that, when repeated, completely describes the three-dimensional pattern of the atoms of a crystal. Geometrically, it can be shown that there are only seven unit cell shapes, or ***crystal systems***, that can be stacked together to fill three-dimensional space. The seven systems, are ***cubic***, ***tetragonal***, ***orthorhombic***, ***rhombohedral***, ***hexagonal***, ***monoclinic***, and ***triclinic***. The various systems are distinguished from one another by the lengths of the unit cell edges and the angles between the edges, collectively known as the ***lattice parameters*** or ***lattice constants*** (a, b, c, α, β, and γ in Figure 1.4).

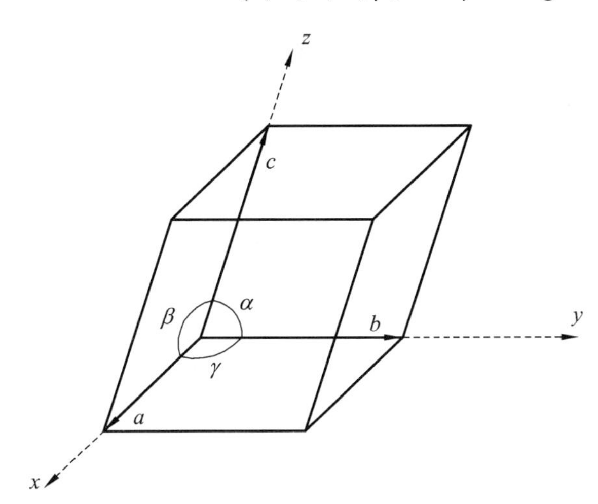

Figure 1.4 Definition of a coordinate system for crystal structures

It is useful to think of the crystal systems as the shape of the "bricks" that make up a solid. For example, the bricks can be cubes, hexagons, ***parallelepipeds***, etc. And while the shape of the bricks is a very important descriptor of a crystal structure, it is insufficient. In addition to the shape of the brick, it is important to know the symmetry of the lattice pattern within each brick as well as the actual location of the atoms on these lattice sites. Only then would the description be complete.

It turns out that if one considers only the symmetry within each unit cell, the number of possible ***permutations*** is limited to 14. The 14 arrangements, shown in Table 1.1, are also known as the Bravais lattices. A lattice can be defined as an indefinitely extending arrangement of points, each of which is surrounded by an identical grouping of neighboring points. To carry the brick brick analogy a little further, the ***Bravais lattice*** represents the symmetry of the pattern found on the bricks. Finally, to describe the atomic arrangement, one must describe the symmetry of the basis, defined as the atom or grouping of atoms located at each lattice site. When the basis is added to the lattices, the total number of possibilities increases to 32 ***point groups.***

Table 1.1 Geometric characteristics of the 7 crystal systems and 14 Bravais lattices

Crystal Structure	**Lattice Parameters**	**Interaxial Angles**	
Cubic	$a=b=c$	$\alpha=\beta=\gamma=90°$	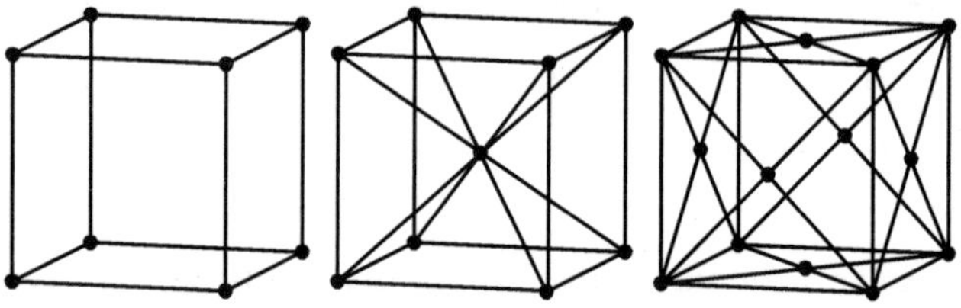 Simple, Body-centered, Face-centered
Orthorhombic	$a\neq b\neq c$	$\alpha=\beta=\gamma=90°$	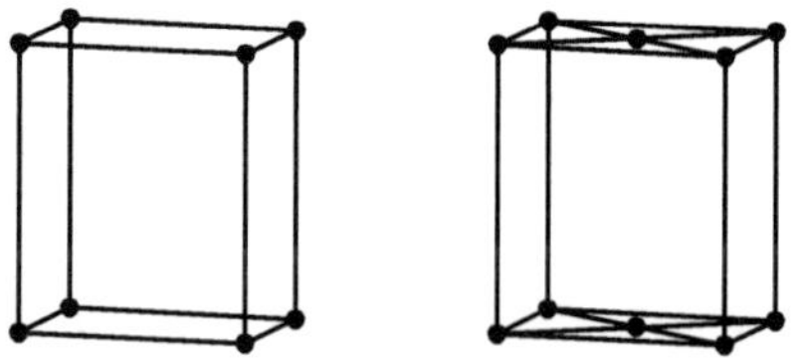 Simple, Base-centered, 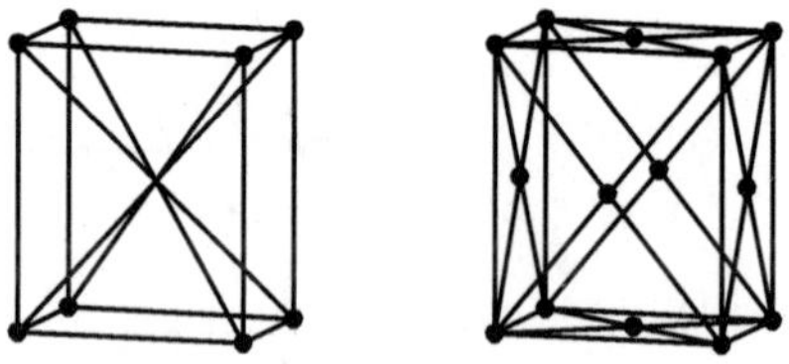 Body-centered, Face-centered
Rhombohedral	$a=b=c$	$\alpha=\beta=\gamma\neq 90°$, $<120°$	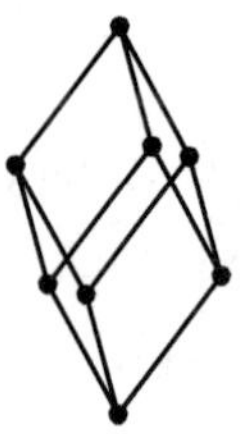

Continued Table 1.1

Crystal Structure	Lattice Parameters	Interaxial Angles	
Tetragonal	$a=b\neq c$	$\alpha=\beta=\gamma=90°$	Simple; Body-centered
Monoclinic	$a\neq b\neq c$	$\alpha=\gamma=90°$, $\beta\neq 90°$	Simple; Base-centered
Triclinic	$a\neq b\neq c$	$\alpha=\beta=\gamma\neq 90°$	
Hexagonal	$a=b$, $a\neq c$	$\alpha=\beta=90°$, $\gamma=120°$	

Anisotropy

The properties of crystals are different in various crystallographic directions, which are associated with an ordered arrangement of atoms (ions, molecules) in space. The phenomenon is called ***anisotropy***.

The properties of crystal are determined by interactions of atoms. In crystals, the spacings between atoms are different in various crystallographic directions, because of which their properties are also different.

Virtually all properties of crystals are anisotropic. The phenomenon is however more pronounced in crystals with structures of a poor symmetry.

Anisotropy of properties is mainly observed in single grown crystals. Natural crystalline solids are mostly polycrystals, i.e. they consist of a plurality of differently oriented fine crystals and exhibit no anisotropy, since the mean statistic spacings between atoms are essentially the same in all directions. In that connection, polycrystalline solids are considered to be quasi-isotropic. After plastic working of a polycrystal, crystallographic planes of the same index may turn out to be oriented in parallel. Such polycrystals are called textured and , like single crystals, they are anisotropic.

New Words and Expressions

long-range order ['lɔ:ŋ'reɪndʒ'ɔ:də] 长程有序
short-range order ['ʃɔ:t'reɪndʒ'ɔ:də] 短程有序
noncrystalline ['nɒn'krɪstəlaɪn] *adj.* 非结晶的
melting point ['meltiŋ pɔint] *n.* 熔点
solidification [səˌlɪdɪfɪ'keɪʃn] *n.* 凝固
inorganic [ˌɪnɔ:'gænɪk] *adj.* 无机的
nylon fibre ['nailɔn 'faibə] 尼龙纤维
spatial arrangement ['speiʃəl ə'reindʒmənt] 空间排列
microvoid [maɪkrəʊ'vɔɪd] *n.* 微空洞，微孔
orientation [ˌɔ:riən'teɪʃn] *n.* 方向，方位，倾向性
crystallographic [ˌkrɪstələ'græfɪk] *adj.* 结晶学的
inhomogeneity ['ɪnˌhəʊmədʒɪ'ni:ɪtɪ] *n.* 不均一（性），多相（性），不同族
unit cell ['ju:nit sel] 晶胞
crystal system ['kristəl 'sistəm] 晶系
cubic ['kju:bɪk] *adj.* 立方的
tetragonal [te'trægənəl] *adj.* 四方的
orthorhombic [ɔ:θə'rɒmbɪk] *adj.* 正交的
rhombohedral [rɒmbəʊ'hi:drəl] *adj.* 斜方六面体的，菱形的
hexagonal [heks'ægənl] *adj.* 六角形的
monoclinic [ˌmɒnə'klɪnɪk] *adj.* 单斜的
triclinic [traɪ'klɪnɪk] *adj.* 三斜的
lattice parameter ['lætɪs pə'ræmitə] 晶格（点阵）参数
lattice constant ['lætɪs'kɔnstənt] 晶格（点阵）常数
parallelepiped [ˌpærəle'lepɪped] *n.* 平行六面体
permutation [ˌpɜ:mju'teɪʃn] *n.* 序列，排列
Bravais lattice 布拉维点阵
point group [pɔint gru:p] 点阵，点群
anisotropy [ə'nɪsətrəpɪ] *n.* 各向异性

Reading Material

Types of Bonds in Materials

Electronegativity is a very useful quantity to help categorize bonds, because it provides a measure of the excess binding energy between atoms A and B, Δ_{A-B} (in $kJ \cdot mol^{-1}$):

$$\Delta_{A-B} = 96.5(\chi_A - \chi_B)^2$$

The excess ***binding energy***, in turn, is related to a measurable quantity, namely the bond dissociation energy between two atoms, DE_{ij}:

$$\Delta_{A-B} = DE_{AB} - [(DE_{AA})(DE_{BB})]^{1/2}$$

The bond ***dissociation energy*** is the energy required to separate two bonded atoms. The greater the electronegativity difference, the greater the excess binding energy. These quantities give us a method of characterizing bond types. More importantly, they relate to important physical properties, such as melting point (see Table 1.2). First, let us review the bond types and characteristics, then describe each in more detail.

Table 1.2 Examples of substances with different types of interatomic bonding

Type of Bond	Substance	Bond Energy /$kJ \cdot mol^{-1}$	Melting Point/°C	Characteristics
Ionic	$CaCl_2$	651	646	Low electrical conductivity, transparent, brittle, high melting point
	$NaCl_2$	768	801	
	LiF	1008	870	
	CuF_2	2591	1360	
	Al_2O_3	15192	3500	
Covalent	Ge	315	958	Low electrical conductivity, very hard, very high melting point
	GaAs	~315	1238	
	Si	353	1420	
	SiC	1188	2600	
	Diamond	714	3550	
Metallic	Na	109	97.5	High electrical and thermal conductivity, easily deformable, opaque
	Al	311	660	
	Cu	340	1083	
	Fe	407	1535	
	W	844	3370	

Continued Table 1.2

Type of Bond	Substance	Bond Energy /kJ · mol^{-1}	Melting Point/°C	Characteristics
Van der Waals	Ne	2.5	− 248.7	Weak binding, low melting and boiling points, very compressible
	Ar	7.6	− 189.4	
	CH_4	10	− 184	
	Kr	12	− 157	
	Cl_2	31	− 103	
Hydrogen bonding	HF	29	− 92	Higher melting point than van der Waals bonding, tendency to form groups of many molecules
	H_2O	50	0	

Primary Bonds

Primary bonds, also known as "strong bonds," are created when there is direct interaction of electrons between two or more atoms, either through transfer or as a result of sharing. The more electrons per atom that take place in this process, the higher the bond "order" (e.g., single, double, or triple bond) and the stronger the connection between atoms. There are four general categories of primary bonds: ionic, covalent, polar covalent, and metallic.

An ionic bond, also called a ***heteropolar*** bond, results when electrons are transferred from the more ***electropositive*** atom to the more electronegative atom, as in ***sodium chloride***, NaCl. Ionic bonds usually result when the electronegativity difference between two atoms in a diatomic molecule is greater than about 2.0. Because of the large discrepancy in electronegativities, one atom will generally gain an electron, while the other atom in a diatomic molecule will lose an electron. Both atoms tend to be "satisfied" with this arrangement because they oftentimes end up with ***noble gas*** electron configurations—that is, full electronic orbitals. The classic example of an ionic bond is NaCl, but CaF_2 and MgO are also examples of molecules in which ionic bonding dominates.

A covalent bond, or ***homopolar*** bond, arises when electrons are shared between two atoms (e.g., H—H). This means that a binding electron in a covalent diatomic molecule such as H_2 has equal likelihood of being found around either ***hydrogen*** atom. Covalent bonds are typically found in homonuclear diatomics such as O_2 and N_2, though the atoms need not be the same to have similar electronegativities. Electronegativity differences of less than about 0.4 characterize covalent bonds. For two atoms with an electronegativity difference of between 0.4 and 2.0, a polar covalent bond is formed—one that is neither truly ionic nor totally covalent. An example of a polar covalent bond can be found in the molecule ***hydrogen fluoride***, HF. Though there is significant sharing of the electrons, some charge distribution exists that results in a polar or partial ionic character to the bond. The percent ionic character of the bond can again be related to the electronegativities of the individual atoms:

$$\text{ionic character}(\%) = 100\{1-\exp[-0.25(\chi_A-\chi_B)^2]\}$$

The larger the electronegativity difference, the more ionic character the bond has. Of course, if the electronegativity difference is greater than about 2.0, we know that an ionic bond should result.

Finally, a special type of primary bond known as a metallic bond is found in an assembly of ***homonuclear*** atoms, such as copper or sodium. Here the bonding electrons become "decentralized" and are shared by the core of positive ***nuclei***. Metallic bonds occur when elements of low electronegativity (usually found in the lower left region of the periodic table) bond with each other to form a class of materials we call metals. Metals tend to have common characteristics such as ***ductility***, ***luster***, and high thermal and ***electrical conductivity***. All of these characteristics can to some degree be accounted for by the nature of the metallic bond. The model of a metallic bond, first proposed by Lorentz, consists of an assembly of positively charged ion cores surrounded by free electrons or an "electron gas." We will see later on, when we describe intermolecular forces and bonding, that the electron cloud does indeed have "structure" in the quantum mechanical sense, which accounts nicely for the observed electrical properties of these materials.

Secondary Bonds

Secondary bonds, or weak bonds, occur due to indirect interaction of electrons in adjacent atoms or molecules. There are three main types of secondary bonding: hydrogen bonding, dipole-dipole interactions, and van der Waals forces. The latter, named after the famous Dutch physicist who first described them, arise due to momentary electric dipoles (regions of positive and negative charge) that can occur in all atoms and molecules due to statistical variations in the charge density. These intermolecular forces are common, but very weak, and are found in ***inert gases*** where other types of bonding do not exist.

Hydrogen bonding is the attraction between hydrogen in a highly polar molecule and the electronegative atom in another polar molecule. In the water molecule, oxygen draws much of the electron density around it, creating positively charged centers at the two hydrogen atoms. These positively charged hydrogen atoms can interact with the negative center around the oxygen in adjacent water molecules. Although this type of bonding is of the same order of magnitude in strength as van der Waals bonding, it can have a profound influence on the properties of a material, such as boiling and melting points. In addition to having important chemical and physical implications, hydrogen bonding plays an important role in many biological and environmental phenomena. It is responsible for causing ice to be less dense than water (how many other substances do you know that are less dense in the solid state than in the liquid state?), an occurrence that allows fish to survive at the bottom of frozen lakes.

Finally, some molecules possess permanent charge separations, or dipoles, such as are found in water. The general case for the interaction of any positive dipole with a negative dipole is called dipole-dipole interaction. Hydrogen bonding can be thought of as a specific type of dipole-dipole interaction. A dipolar molecule like ***ammonia***, NH_3, is able to dissolve other polar molecules, like water, due to dipole-dipole interactions. In the case of NaCl in water, the dipole-dipole interactions are so strong as to break the intermolecular forces within the molecular solid.

New Words and Expressions

electronegativity [elektrəʊnɪgə'tɪvɪtɪ] *n.* 负电性，阴电性

binding energy ['baɪndɪŋ'enədʒi] 结合能

dissociation energy [diˌsəuʃi'eiʃən 'enədʒi] 离解能

heteropolar [ˌhetərə'pəʊlə] *adj.* 异极的

electropositive [ɪˌlektrəʊ'pɒzətɪv] *adj.* 正电性的，阳电性的

sodium chloride ['səʊdiːəm'klɔːraid] 氯化钠

noble gas ['nəubl gæs] *n.* 惰性气体

homopolar [ˌhəʊmə'pəʊlə] *adj.* 同极的，单极的

homonuclear [hɒmə'njuːklɪə(r)] *adj.* 同核的，共核的

hydrogen fluoride ['haidrədʒən'flʊərˌaɪd] 氟化氢

nuclei ['njuːklɪaɪ] *n.* 核，核心，原子核（nucleus 的复数形）

ductility [dʌk'tɪlɪtɪ] *n.* 延展性，柔软性，韧性，塑性

luster ['lʌstə] *n.* 光泽，光彩

electrical conductivity [i'lektrikəl,kɔndʌk'tɪvɪti:] 电导率

inert gas [ɪn'ɜ:t gæs] *n.* 惰性气体

ammonia [ə'məʊniə] *n.* 氨

Lesson 3 Structure-property-processing Relationship

We are interested in producing a component that has the proper shape and properties, permitting the component to perform its task for its expected lifetime. The materials engineer meets this requirement by taking advantage of a complex three-part relationship between the internal structure of the material, the processing of the material, and the final properties of the material (Figure 1.5). When the materials engineer changes one of these three aspects of the relationship, either or both of the others also change. We must therefore determine how the three aspects interrelate in order to finally produce the required product.

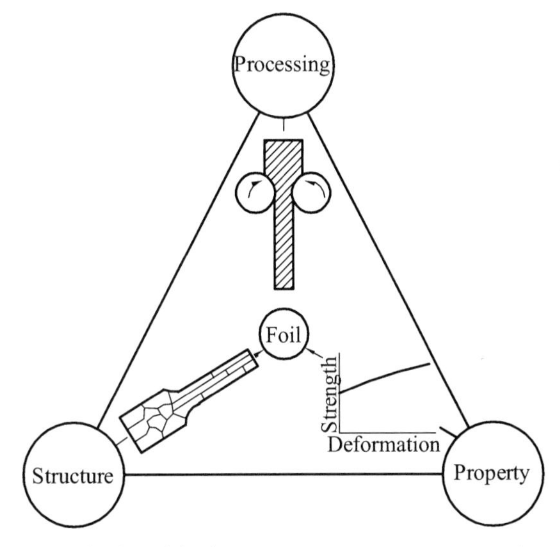

Figure 1.5 The three-part relationship between structure, property and processing method

Structure

The structure of a material can be considered on several levels, all of which influence the final behavior of the product. At the finest level is the structure of the individual atoms that compose the material. The arrangement of the electrons surrounding the nucleus of the atoms significantly affects electrical, magnetic, thermal, and optical behavior and may also influence ***corrosion resistance***. Furthermore, the electronic arrangement influences how the atoms are bonded to one another and helps determine the type of material—metal, ceramic, or polymer.

At the next level, the arrangement of the atoms in space is considered. Metals, many ceramics, and some polymers have a very regular atomic arrangement, or crystal structure. The crystal structure influences the ***mechanical properties*** of metals such as ductility, strength, and ***shock resistance***. Other ceramic materials and most polymers have no orderly atomic arrangement—these amorphous or glassy materials behave much differently from crystalline materials. For instance, glassy ***polyethylene*** is transparent while crystalline polyethylene is translucent. ***Defects*** in this atomic arrangement exist and may be controlled to produce profound changes in properties.

A grain structure is found in most metals, some ceramics, and occasionally in polymers. Between the grains, the atomic arrangement changes its orientation and thus influences properties. The size and shape of the grains play a key role at this level.

Finally, in most materials, more than one phase is present, with each phase having its unique atomic arrangement and properties. Control of the type, size, distribution, and amount of these phases within the main body of the material provides an additional way to control properties.

Properties

A material exhibits a set of properties, which define its behavior. A property of a material is determined by analyzing the reaction of the material to some outside influence, generally by means of a ***normalized standard test***. According to the type of outside influence, three categories of properties are recognized.

Mechanical properties, which reflect the behavior of materials, deformed by a set of forces.

Physical properties, which describe the behavior of materials subjected to the action of temperature, electric or magnetic fields, or light.

Chemical properties, which characterize the behavior of material in reactive environment.

We can consider the properties of a material in two categories—mechanical and physical here (Table 1.3).

Table 1.3 Typical examples of the properties of materials

Mechanical Properties	Physical Properties
Creep	**Chemical**
Creep rate	Corrosion
Stress-rupture properties	Refining
Ductility	**Density**
% Elongation	**Electrical**
% Reduction in area	Conductivity
Fatigue	Dielectric (insulation)
Endurance limit	***Ferroelectricity***
Fatigue life	***Piezoelectricity***
Hardness	**Magnetic**
Scratch resistance	***Ferrimagnetic***
Wear rates	***Ferromagnetic***
Impact	***Paramagnetic***
Absorbed energy	**Optical**
Toughness	Absorption
Transition temperature	Color
Strength	***Diffraction***
Modulus of elasticity	Lasing action
Tensile strength	Photoconduction
Yield strength	***Reflection***
	Refraction
	Transmission
	Thermal
	Heat capacity
	Thermal conductivity
	Thermal expansion

The mechanical properties describe how the material responds to an applied force or stress. Stress is defined as the force divided by the cross-sectional area on which the force acts. The most common mechanical properties are the strength, ductility, and ***stiffness*** of the material. However, we are often interested in how the material behaves when it is exposed to a sudden intense blow (impact), continually cycled through an alternating force (fatigue), exposed to high temperatures (stability), or subjected to abrasive conditions (wear). The mechanical properties not only determine how well the material performs in service, but also determine the ease with which the material can be formed into a useful shape. A metal part formed by forging must withstand the rapid application of a force without breaking and have a high enough ductility to deform to the proper shape. Often small changes in the structure have a profound effect on the mechanical properties of the material.

Physical properties include electrical, magnetic, optical, thermal, elastic, and chemical behavior. The physical properties depend both on structure and processing of the material. Even tiny changes in the composition cause a profound change in the electrical conductivity of many semiconducting metals and ceramics.

High firing temperatures may greatly reduce the thermal insulation characteristics of ceramic brick. Small amounts of impurities change the color of a glass or polymer.

Processing

Materials processing produces the desired shape of a component from the initial formless material. Metals can be processed by pouring liquid metal into a mold (***casting***), joining individual pieces of metal (***welding***, ***brazing***, ***soldering***, adhesive bonding), forming the solid metal into useful shapes using high pressures (forging, ***drawing***, ***extrusion***, rolling, bending), compacting tiny metal powder particles into a solid mass (***powder metallurgy***), or removing excess materials (***machining***). Similarly, ceramic materials can be formed into shapes by related processes such as casting, forming, extrusion, or compaction, often while wet, and heat treatment at high temperatures to drive off the fluids and to bond the individual constituents together. Polymers are produced by injection of softened plastic into molds (much like casting), drawing and forming. Often a material is heat treated at some temperature below its melting temperature to effect a desired change in structure. The type of processing we use depends, at least partly, on the properties, and thus the structure, of the material.

Structure-property-processing Interaction

The processing of a material affects the structure. The structure of a copper bar is very different if it is produced by casting rather than forming. The shape, size, and orientation of the grains may be different, the cast structure may contain voids due to shrinkage or gas bubbles, and nonmetallic particles (inclusions) may be trapped within the structure. The formed material may contain elongated nonmetallic particles and internal defects in the atomic arrangement. Thus, the structure and consequently the final properties of the casting are very different from those of the formed product.

On the other hand, the original structure and properties determine how we can process the material to produce a desired shape. A casting containing large ***shrinkage*** voids may crack during a subsequent processing step. Alloys that have been strengthened by introducing imperfections in the structure also become brittle and fail during forming. Elongated grains in a metal may lead to nonuniform shapes when subsequently formed. Thermosetting polymers cannot be formed, while thermoplastic polymers are easily formed.

Summary

The procedure of selecting a material, processing the material into a useful shape, and obtaining the needed properties is a complicated process involving knowledge of the structure-property-processing relationship. The remainder of this text is intended to introduce the students to the wide variety of materials available. As we do so, we will come to understand the fundamentals of the structure of material, how the structure affects the behavior of the materials and the role that processing and the environment play in shaping the relationship between structure and properties.

New Words and Expressions

corrosion resistance [kə'rəʊʒən ri'zistəns] 耐腐蚀性
mechanical property [mi'kænikəl 'prɔpəti] 力学性能
shock resistance [ʃɔk ri'zistəns] 冲击强度，抗冲击性
polyethylene [ˌpɒli'eθəli:n] *n.* 聚乙烯
defect ['di:fekt] *n.* 瑕疵，缺点
normalized standard test 标准试验
creep [kri:p] *n.* 蠕变
fatigue [fə'ti:g] *n.* 疲劳
hardness [hɑ:dnəs] *n.* 硬度
scratch resistance [skrætʃ ri'zistəns] 抗划伤性
transition temperature [træn'ziʃən 'tempəritʃə] 脆性转变温度
modulus of elasticity ['mɔdjuləs ɔv ɪlæ'stɪsɪti:] 弹性模量
tensile strength ['tensəl streŋθ] 抗拉强度
yield strength [ji:ld streŋθ] 屈服强度
ferroelectricity [ferəʊɪ'lektrɪsɪtɪ] *n.* 铁电性
piezoelectricity [paɪˌi:zəʊɪˌlek'trɪsɪtɪ] *n.* 压电性
ferrimagnetic [ferɪmæg'netɪk] *adj.* 亚铁磁的
ferromagnetic [ˌferəʊmæg'netɪk] *adj.* 铁磁性的

paramagnetic [ˌpærəmæg'netɪk] *adj.* 顺磁性的
diffraction [dɪ'frækʃn] *n.* 衍射
reflection [rɪ'flekʃn] *n.* 反射
refraction [rɪ'frækʃn] *n.* 折射
heat capacity [hiːt kə'pæsiti] 热容
thermal conductivity ['θəːməl ˌkɔndʌk'tɪvitiː] 导热性
thermal expansion ['θəːməl iks'pænʃən] 热膨胀
stiffness [stɪfnəs] *n.* 刚度
forging ['fɔːdʒɪŋ] *n.* 铸造
casting ['kɑːstɪŋ] *n.* 铸造，铸件
welding [weldɪŋ] *n.* 焊接法，定拉焊接
brazing [breɪzɪŋ] *n.* 铜焊
soldering ['sɒldərɪŋ] *n.* 锡焊
drawing ['drɔːɪŋ] *n.* 拉拔，拔丝
extrusion [ɪk'struːʒn] *n.* 挤出
machining [mə'ʃiːnɪŋ] *v.* 加工
shrinkage ['ʃrɪŋkɪdʒ] *n.* 收缩

Reading Material

Defects in Crystals

The structure of real crystals differs from that of ideal ones. Real crystals always have certain defects, and therefore, the arrangement of atoms in the volume of a crystal is far from being perfectly regular.

It is distinguished between point, linear, surface and volume defects. The dimensions of a point defect are close to those of an interatomic space (zero dimension). With linear defects, also known as ***dislocations***, are one dimensional, their length is several orders of magnitude greater than the width. Surface defects have a small depth, while their width and length may be several orders larger. Volume defects (pores and cracks) may have substantial dimensions in all measurements. These defects may occur individually or in combination.

Point Defects

The simplest point defects are vacancies, interstitial atoms of the base substance (interstitials), and foreign interstitial atoms (Figure 1.6).

(a) Vacancy

(b) Interstitial atom

(c) Interstitial impurity atom

Figure 1.6 Point defects in a crystal lattice

A vacancy is an empty (unoccupied) site of a crystal lattice; an interstitial atom is an atom transferred from a site into an interstitial position.

Vacancies and interstitial atoms can appear in crystals at any temperature above the absolute zero owing to thermal oscillations of atoms. Each temperature has a corresponding equilibrium concentration of vacancies and interstitial atoms. For instance, copper can contain 10^{-13} at.% of vacancies at a temperature of 20-25 °C and so many as 0.01 at.% near the melting point (one vacancy per 10^4 atoms).

Supersaturation with point defects can be achieved by sharp cooling after high-temperature heating, on plastic deformation or neutron irradiation. In the last case, the concentration of vacancies is the same as that of interstitial atoms: atoms knocked off from sites of the lattice become interstitial atoms, whereas the sites they leave become vacancies. In the course of time, the surplus of vacancies above the equilibrium concentration on free surfaces of a crystal, pores, grain boundaries and other lattice defects is annihilated. The points where vacancies are annihilated are called ***vacancy sinks***. Annihilation of vacancies can be attributed to their high mobility in the lattice. An atom near a vacancy can occupy that vacancy and leave empty its site which then will be occupied by another atom.

At higher temperatures, vacancies have a higher concentration and can move from one site to another more frequently. Vacancies are the most important kind of points defects; they accelerate all processes associated with displacements of atoms: diffusion, powder sintering, etc.

In ionic and covalent crystals, vacancies are electrically active and may serve both as donors and as acceptors. This can create a certain prevailing type of conductance in crystals. In ionic crystals, their neutrality is retained owing to the formation of a couple of point defects: vacancy-ion which have electric charges of opposite signs.

All kinds of point defects distort the crystal lattice and have a certain influence on the physical properties. In commercially pure metals, point defects increase the electric resistance and have almost no effect on the mechanical properties. Only at high concentrations of defects in irradiated metals, the ductility and other properties are reduced noticeably.

Linear Defects

The most important kinds of linear defects are edge and screw dislocations (Figure 1.7).

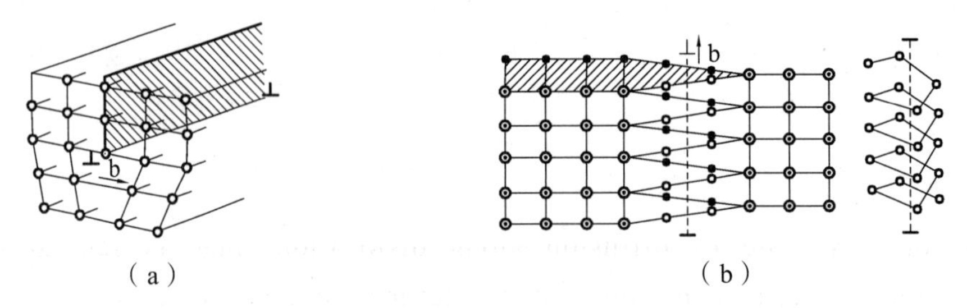

Figure 1.7 Scheme of (a) edge dislocation and (b) screw dislocation

An edge dislocation in its cross-section is essentially the edge of an "extra" half-plane in the crystal lattice [Figure 1.8 (a)]. The lattice around dislocations is elastically distorted.

The criterion of distortion is what is called the Burgers vector. It can be determined if a closed contour is drawn around a zone in an ideal crystal by passing from one site to another [Figure 1.9 (a)] and then the procedure is repeated for a zone in a real crystal containing a dislocation. As may be seen from Figure 1.9 (b), the contour described in the real crystal turns out to be unclosed. The vector required for closing the contour is the Burgers vector. The Burgers vector of an edge dislocation is equal to the interatomic space and ***perpendicular*** to the dislocation line; for a screw dislocation, it is parallel to the dislocation line.

Figure 1.8 Dislocations in annealed Fe + 3% Al alloy

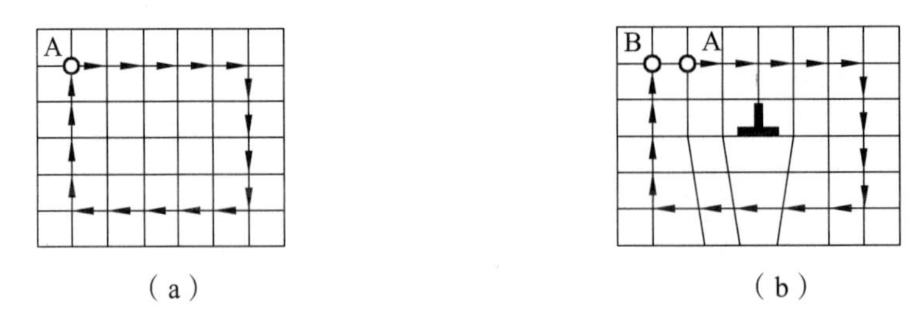

Figure 1.9 Determination of Burgers vector (BA)

The total length of all dislocation lines in a unit of volume is called the dislocation density. It may be equal to 10^4-10^5 cm^{-2} in semiconductor crystals and 10^6-10^8 cm^{-2} in annealed metals. After cold plastic deformation, the dislocation density may rise up to 10^{11}-10^{12} cm^{-2}. Attempts to raise the dislocation density above 10^{12} cm^{-2} end quickly in cracking and failure of the metal. Dislocations appear on crystallization; their density may then be quite high and they influence substantially the properties of materials. Along with other defects, dislocations participate in phase transformations and ***recrystallization*** and may serve as nuclei for precipitation of a secondary phase from solid solution. The rate of diffusion along dislocation lines is several orders of magnitude greater than that through a crystal lattice without defects. Dislocations serve as places for concentration of impurity atoms, especially of interstitial impurities, since this decreases lattice distortions. Impurity atoms can concentrate around dislocations and form what is called Cottrell atmospheres which impede dislocation movement and strengthen the metal.

The effect of dislocations is especially pronounced on the strength of crystals. The experimentally measured yield strength of metals turns out to be only one-thousandth of its theoretical value, the loss being mainly attributed to the effect of mobile dislocations. By increasing substantially the dislocation density and decreasing the dislocation mobility, the strength of a metal can be raised several times compared with its strength in the annealed state. Faultless pieces of metals (in particular, long and thin "whiskers" obtained by crystallization from the gaseous phase) exhibit a strength approaching the theoretical value (Figure 1.10).

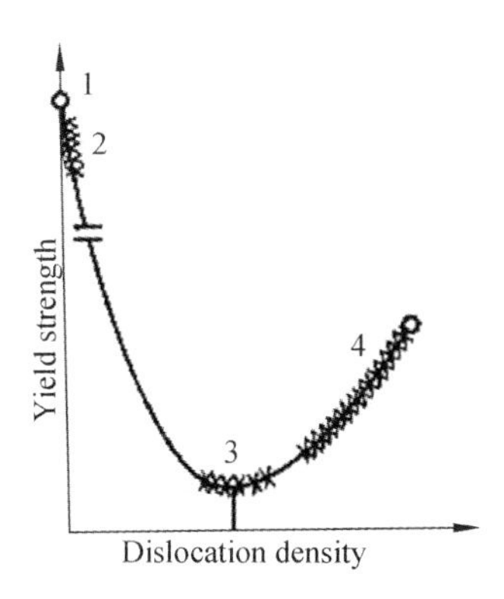

Figure 1.10 Effect dislocation density on yield strength

1—a perfect faultless crystal; 2—faultless "whisker" crystals; 3—annealed metals; 4—metals with an elevated stacking

In semiconductors, dislocations can influence the electric and other properties, in particular, decrease the electric resistance and shorten the life of carriers. The role of dislocation is especially important in microelectronics where thin filmlike crystals are used and dislocations can play the part of narrow conducting channels through which impurity atoms can move easily.

Surface Defects

The most important kinds of surface defects are high-angle and twin boundaries.

Engineering materials may be either polycrystalline or single-crystal type. A polycrystalline alloy contains an enormous quantity of fine grains. The lattices of adjacent grains are oriented at random and differently (Figure 1.11) and a boundary between any two grains is essentially a transition layer of thickness of 1-5 nm. This layer may have a disordered arrangement of atoms, dislocation clusters, and an elevated concentration of impurities.

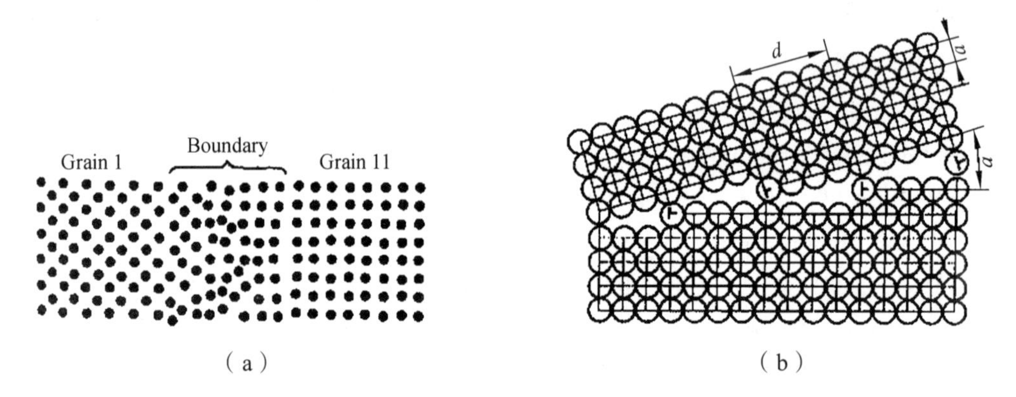

Figure 1.11 Schemes of (a) high-angle and (b) low-angle boundaries

High-angle boundaries are the boundaries between grains, since the corresponding crystallographic directions in adjacent grains make angles of ten of degrees with one another [Figure 1.11 (a)]. Each grain in turn consists of subgrains or blocks.

A subgrain is a portion of a crystal of a relatively regular structure. Subgrain boundaries are formed by walls of dislocations which divide a grain into a number of subgrains or blocks [Figure 1.11 (b)]. Angles of misorientation between adjacent subgrains are not large (not more than 5°), so that their boundaries are termed "low-angle". Low-angle boundaries can also serve as places of accumulation of impurities.

Surface defects influence the mechanical and physical properties of materials. Of especially large importance are grain boundaries. The yield strength σ_y is associated with the grain size d by the relationship $\sigma_y = \sigma_0 + kd^{-1/2}$, where σ_0 and k are constants for a given material. With finer grain, a material has higher yield strength and toughness and is less susceptible to brittle failure. The size of subgrains has a similar, though less strong, effect on mechanical properties.

Diffusion along boundaries of grains and subgrains occurs many times quicker than in other directions in a crystal, especially on heating. Interactions of defects, their displacements in crystals, and variations of their concentrations—all these factors can change the properties of metals and are of high practical significance.

New Words and Expressions

dislocation [ˌdɪslə'keɪʃn] *n.* 错位
supersaturation [ˌsuːpəsætʃə'reɪʃən] *n.* 过饱和
vacancy sink ['veɪkənsiː siŋk] 空位聚居
anneal [ə'niːl] *n.* 退火；*vt.* 使退火
perpendicular [ˌpɜːpən'dɪkjələ(r)] *adj.* 垂直的，正交的；*n.* 垂直线，直立
recrystallization *n.* 再结晶，重结晶

UNIT 2 METALLIC MATERIALS

Lesson 4 Introduction to Metallic Materials

What is a Metal?

The key feature that distinguishes metals from non-metals is their bonding. Metallic materials have free electrons. In the case of pure metals, the outermost layer of electrons is not bound to any given atom, instead these electrons are free to roam from atom to atom. Thus, the structure of metallic material can be thought of a consisting of positive centers (or ions) sitting in a "gas" of free-electrons.

The existence of this free electron gas has a number of profound consequences for the properties of metallic materials. For example, one of the most important features of metallic materials is that freely moving electrons can conduct electricity and so metallic materials tend to be good electrical conductors. Some metals more closely resemble the idealized picture of free electrons than others. Consequently, some metals are better conductors of electricity than others; for example, copper is a more efficient electrical conductor than tin.

Electrical conductivity is such an important characteristic of metals that conductivity is sometimes used to distinguish metals from non-metals. The problem with using conductivity to distinguish metals from non-metals is that this approach is somewhat arbitrary. For example, ***graphite*** is a form of carbon which has quite a high electrical conductivity, but the bonding of carbon atoms in graphite is very different from that of atoms in a metal. Therefore, it would be quite misleading to describe graphite as a metal. Note: the way in which atoms are arranged in a structure is just as important as the nature of the atoms themselves in determining the extent of electrical conductivity. Both diamond and graphite are made up of pure carbon, but diamond is a very good electrical insulator, rather than an electrical conductor like graphite.

What is an Alloy?

An alloy consists of a mixture of a pure metal and one or more other elements. Often, these other elements will be metals. For example, brass is an alloy of copper and zinc. In other cases, a metal will be alloyed with a non-metal. The most important example of alloying involving addition of a non-metal would be (plain-carbon) steels, which consist of iron alloyed with carbon.

Alloys are usually less ***malleable*** and ductile than pure metals and the tend to have lower melting points. They do, however, have other properties which make them more useful than pure metals. An alloy is made by melting the different metals in the alloy together. The amounts of each metal are usually quite important.

The Two Classes of Metallic Materials

Metallic materials have been conveniently grouped into two classes: ***ferrous metals*** and ***nonferrous metals***.

Ferrous Metals

Iron and its many alloys, including cast irons and a nearly limitless variety of steels, comprise the ferrous metals group. ***Wrought iron*** is a commercial iron consisting of ***slag*** (***iron silicate***) fibers entrained in a ferrite matrix. It contains approximately 2% slag but very little carbon, and is easily shaped by hot-forming operations such as forging. ***Ingot iron*** contains about 0.1% impurities including 0.01% carbon. It is used in applications where high ductility or corrosion resistance is needed. ***Electrolytic iron***, about 99.99% pure, is used mostly for research.

Pig iron is iron that is tapped from the base of the blast furnace and contains over 4% carbon plus other elements (impurities). It is converted into ***gray cast iron*** by heating, along with scrap, in a vertical furnace called a cupola. For steelmaking, it is refined into steel in the basic oxygen furnace (BOF) or electric furnace. Even with the wide acceptance of aluminum and polymeric materials, the iron-based alloys dominate all other materials in the weight consumed annually for manufactured products. Ten times more iron (mainly in the form of steel) is used than all other metals combined.

Only limited amounts of pure iron, usually ingot iron or iron powders, are used in modern industry. Steels of iron and alloying elements, such as carbon, silicon, ***nickel***, ***chromium***, and ***manganese***, are more widely used. A plain-carbon steel, which contains carbon, silicon, and manganese, rarely contains more than 1% of an alloying element. A low-alloy steel contains carbon, silicon, and manganese, together with small quantities of nickel, chromium, molybdenum, and other alloying elements that alter the properties of steel. High-alloy steels have higher quantities (more than 5%) of alloying elements.

The uses of some ferrous metals are listed according to their carbon content in Table 2.1.

Table 2.1 Use of ferrous metals by carbon content

Type	Carbon Range/%	Typical Uses
Carbon Steels		
Low	0.05 to 0.30	·For cold formability; ·Wire, nails, rivets, screws; ·Sheet stock for drawing; ·Fenders, pots, pans, welding rods; ·Bars, plates, structural shapes, shafting; ·Forgings, carburized parts, key-stock; ·Free-machining steel.
Medium	0.30 to 0.60	·Heat-treated parts that require moderate strength and high toughness such as bolts, shafting, axles, spline shafts; ·Higher strength, heat-treated parts with moderate toughness such as lock washers, springs, band saw blades, ring gears, valve springs, snap rings.
High	0.60 to 2.0	·Chisels, center punches; ·Music wire, mower blades, leaf springs; ·Hay rake tines, leaf springs, knives, woodworking tools, files, reamers; ·Ball bearings, punches, dies.
Cast Irons		
Gray	2.0 to 4.5	·Machinable castings such as engine blocks, pipe, gears, lathe beds.
White	2.0 to 3.5	·Non-machinable castings such as cast parts for wear resistance.
Malleable castings	2.0 to 3.5	·Produced from white cast iron; ·Machinable castings such as axle and differential housings, crankshafts, camshafts.
Nodular iron (ductile iron)	2.0 to 4.5	·Machinable castings such as pistons, cylinder blocks and heads, wrenches, forming dies.

The properties of ferrous metals are listed in Table 2.2.

Table 2.2 Ferrous metal properties

Element	Common Content	Effects
Carbon	Up to 0.90%	Increases hardness, tensile strength, and responsiveness to heat treatment with corresponding increases in strength and hardness.
	Over 0.90%	Increases hardness and brittleness; over 1.2%, cause loss of malleability.
Manganese	0.50% to 2.0%	Imparts strength and responsiveness to heat treatment; Promotes hardness, uniformity of internal grain structure.
Silicon	Up to 2.50%	Same general effects as manganese.
Sulfur	Up to 0.05%	Maintained below this content to retain malleability at high temperatures, which is reduced with increased content.
	0.05% to 3.0%	Improves machinability.
Phosphorus	Up to 0.05%	Increase strength and corrosion resistance, but is maintained below this content to retain malleability and weldability at room temperature.

Nonferrous Metals

There are a number of nonferrous metals (those that do not contain iron) and alloys that are widely applied in modern products. The radioactive metals uranium, thorium, and plutonium are used as nuclear fuels. Zirconium is an alloying element and is also used in the nuclear field. The light metals aluminum, beryllium, calcium, lithium, magnesium, potassium, titanium, and sodium also have their particular uses. Aluminum, beryllium and titanium are used as structural metals, whereas the remaining light metals are too soft and chemically reactive; these metals are used to extract metals from their ores. Sodium and potassium are used in the nuclear field as coolants. Nickel and lead are versatile metals used in many applications, whereas copper is used primarily for its thermal and electrical conductivity. Cadmium, tin, and zinc are often used in electrical applications and for bearings. Cobalt and manganese are used as alloying elements in ferrous and nonferrous metals. Silver is used as a decorative metal and in brazing alloys, where gold, silver, and platinum are used for electrical contacts and jewelry. Finally, the ***refractory*** metals, those with melting points above 2000 °C, such as columbium, titanium, tungsten, vanadium, and zirconium, are used in applications requiring high strength, hardness, and high temperatures.

New Words and Expressions

graphite ['græfaɪt] *n.* 石墨
malleable ['mæliəbl] *adj.* 有延展性的，可锻的
ferrous metal ['ferəs 'metl] 黑色（铁类）金属
nonferrous metal ['nɔn'ferəs 'metl] 有色金属，非铁金属
wrought iron [rɔːt 'aiən] *n.* 熟铁，锻铁
slag [slæg] *n.* 矿渣，熔渣
ingot iron ['ɪŋgət 'aiən] 锭铁，低碳铁
electrolytic iron [iˌlektrəu'litik 'aiən] 电解铁
pig iron [pig 'aiən] 生铁
gray cast iron [grei kɑːst 'aiən] 灰口铁
nickel ['nɪkl] *n.* 镍，五分镍币
chromium ['krəʊmiəm] *n.* 铬
manganese ['mæŋgəniːz] *n.* 锰
refractory [rɪ'fræktəri] *adj.* 难熔的

Reading Material

Metals with a Memory

Shape-memory alloys are not new; certain brasses, alloy of copper and zinc, and other alloys were known long ago that could alter their shape with a change in temperature. However, some new alloys show this unusual property to a marked extent. Smart metals, the ***shape-memory alloys*** (SMAs) such as ***Nitinol*** (nickle-titanium alloy), have been around for many years and are beginning to find wide application as designers learn more about them. One application is in eyeglass frames. The high-quality, ***shape-memory frames*** made of this "***smart material***" return to their original shape when bent. SMAs are also implanted in "smart structures", and are used as controls for robot fingers and dental braces. Nitinol, that was discovered at the U.S. Naval Ordinance Laboratory (NOL) in 1958 by ***metallurgist*** William Buehler, shows great promise for future applications.

When nitinol is heated to a high temperature, formed to a desired shape while hot, and then quenched, it has a ***martensite structure***. This metal in this state is not hard but instead is unusually soft and ***pliable***. If it is now reformed cold to a new shape, it will still "remember" its original shape. Now, if the soft metal is heated to its particular transition temperature, it will suddenly reshape itself to its original configuration. The transition temperature can be adjusted by the proportions of nickel and titanium, so that small differences of temperature can produce the memory response. A heat engine using nitinol wire loops was developed in 1973 by Ridgway Banks at the Lawrence Laboratory in Berkeley, California. While this engine was running it was noted that nitinol has a "double memory". When the remembered shape is again cooled in water, it begins to "learn" its second shape in the soft condition during repeated cycles as in a heat engine. This phenomenon is known as tow-way memory. Even by the use of a single memory response, a spring can be used to pull the nitinol back to the original position when it is cooled from the transition temperature, since it becomes soft at the lower temperature.

Considerable experimentation is being done on this new metal, and one surprise is that after millions of cycles, nitinol shows no signs of fatigues or wearing out its shape memory response; indeed, it appears to get stronger with use. This metal has almost unlimited possibilities, such as in the areas of space technology, surgical implants, electrical power generation, and automobile engines.

shape-memory alloys　形状记忆合金
nitinol　['nitinɔl]　*n.* 镍钛诺（镍和钛的合金）
shape-memory frames　形状记忆框架
smart material　[smɑːt mə'tiəriəl]　智能材料
metallurgist　[mə'tælədʒɪst]　*n.* 冶金家，冶金学者
quench　[kwentʃ]　*v.* 淬火，熄灭
martensite structure　马氏体结构
pliable　['plaɪəbl]　*adj.* 易曲折的，柔软的

Lesson 5　Crystal Structure of Metals

Bonding of Metals

There are four basic types of bonding arrangements that hold atoms together. They are ionic, covalent, metallic, and van der Waals forces. A molecule may be held together by a combination of several or all of these forces.

The metallic bond is formed among similar metal atoms when some electrons in the valence shell separate from their atoms and exist in a cloud surrounding all the positively charged atoms. These positively charged atoms arrange themselves in a very orderly pattern. The atoms are held together because of their mutual attraction for the negative electron cloud (Figure 2.1). The free movement of electrons accounts for the high electrical and heat conductivity of metals and for their elasticity and plasticity. The metallic bond is very strong.

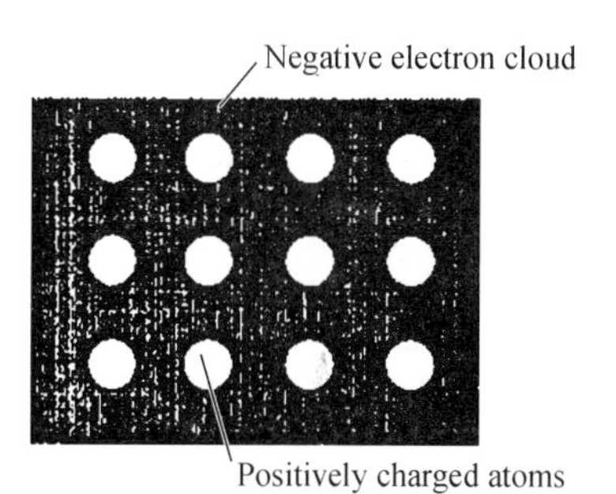

Figure 2.1　Metallic bond

Metals and Nonmetals

Approximately three-quarters of all the elements are considered to be metals. Some of these are ***metalloids*** such as silicon or germanium. Some of the properties that an element must have to be considered a metal are as follows:

(1) ability to donate electrons and form a positive ion;

(2) crystalline structure-grain structure;

(3) high thermal and electrical conductivity;

(4) ability to be ***deformed plastically***;

(5) metallic luster or reflectivity.

Metalloids, such as silicon, possess one or more of these properties, but they are not true metals unless they have all of the characteristics of metal.

Crystalline Unit Structures

When metals solidify from the molten to the solid state, the atoms align themselves in orderly rows, which is an arrangement that is peculiar to that metal. This arrangement is called a space lattice, which can be considered a series of points in space. By means of the study of ***X-ray diffraction***, the space lattice of the crystal has been determined for different metals.

Metals solidify into six main lattice structures:

(1) body-centered cubic (BCC);

(2) face-centered cubic (FCC);

(3) hexagonal close-packed (HCP);

(4) cubic;

(5) body-centered tetragonal;

(6) rhombohedral.

The three most common crystal patterns of unit cells are shown in Figure 2.2.

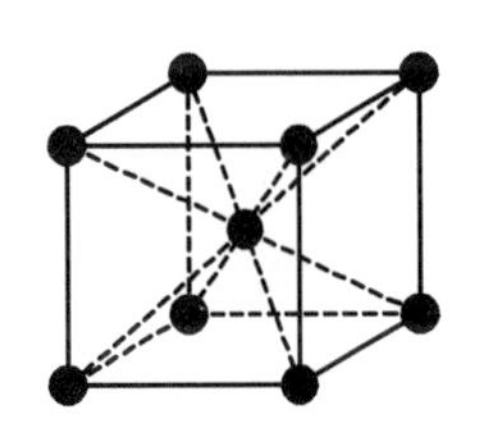
(a) Body-centered cubic

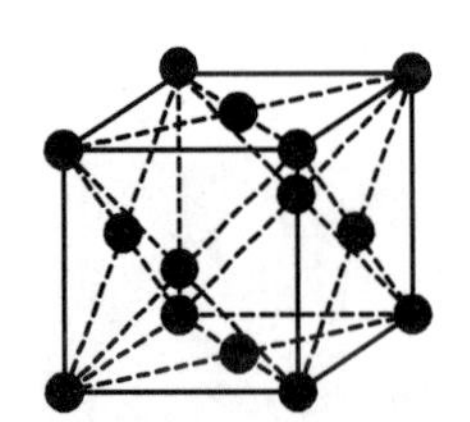
(b) Face-centered cubic

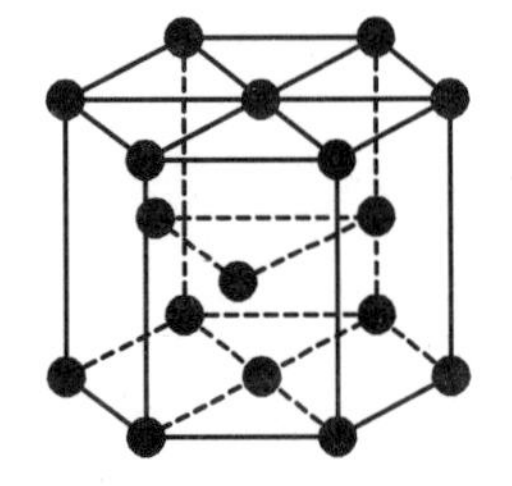
(c) Hexagonal close-packed

Figure 2.2 Three common unit cells of the space lattice

Body-centered Cubic (BCC)

This cubic unit structure is made up of atoms at each corner of the cube and one in the very center. Steel under 723 °C has this arrangement, and it is called ***alpha iron*** or ferrite. Other metals such as chromium, columbium, barium, vanadium, molybdenum, and tungsten crystallize into this lattice structure.

Body-centered cubic metals [see Figure 2.2 (a)] show a lower ductility but a higher yield strength than face-centered cubic metals.

Face-centered Cubic (FCC)

Atoms of calcium, aluminum, copper, lead, nickel, gold, platinum and some other metals arrange themselves with an atom in each corner of the cube and one in the center of each cube face. When steel is above the upper critical temperature, it rearranges its atoms to this FCC structure and is called ***gamma iron*** or ***austenite*** [see Figure 2.2 (b)].

Hexagonal Close-packed (HCP)

The hexagonal close-packed structure [see Figure 2.2 (c)] is found in many of the least common metals. Beryllium, zinc, cobalt, titanium, magnesium and cadmium are examples of metals that crystallize into this structure. Because of the spacing of the lattice structure, rows of atoms do not easily slide over one another in HCP. For this reason, these metals have lower plasticity and ductility than cubic structures.

New Words and Expressions

metalloid ['metlɔɪd] *n.* 非金属
deformed plastically 塑性变形
alpha iron ['ælfə 'aiən] α-铁
gamma iron ['gæmə 'aiən] γ-铁
austenite ['ɔːstəˌnaɪt] *n.* 奥氏体

Reading Material

Crystalline and Amorphous Metals

In a material that is "crystalline" a long range structure exists in which atoms are arranged into unit cells and the unit cells repeat in a regular pattern, forming a lattice. If the lattice extends right out to the edges of a piece of material, the result is a large "single crystal". The most familiar example of this would be ***gemstone***, such as a diamond. However, the nickel-base alloy turbine blades used in aero gas-turbine engines (what are popularly known as jet engines) can also be produced as single crystals. To produce a solid single crystal from a liquid requires that the lattice forms in a uniform fashion. In most cases, however, solidification begins from multiple sites, each of which can produce a different orientation. The result is a "polycrystalline" material consisting of many small crystals (also known as "grains") each of which has the same lattice, but with a ***misorientation*** from grain to grain.

In three dimensions, the grains in a polycrystal are usually found to be ***polygonal***. The ***interfaces*** between the grains ("grain boundaries") have an associated energy, just as do other interfaces (for example the outer surface of a liquid or a solid). For any given type of interface, there will be a certain ***interfacial energy*** per unit area of interface. Hence, since matter tries to adopt the lowest energy condition possible, there is a driving force to minimize the total interfacial area, as this reduces the total interfacial energy of the sample. ***Spherical grains*** would give the lowest surface area to volume ratio, but pacing spheres together can not completely fill space (in other words a sample that solidified by forming spheres would inevitably leave some liquid behind). Hence, the grains do the next best thing and grow as polygons, which have almost as low a surface area to volume ratio as spheres, but unlike spheres can stack together to fill space.

If a material is cooled very rapidly from the liquid state there is not enough time for the material to arrange the solid into a lattice. Instead, a random "amorphous" arrangement is produced and the result is a non-crystalline material. The best known amorphous material is window glass and hence amorphous materials are often referred to as glasses. Metals are usually able to crystallize even at very high cooling rates, but under extreme conditions metallic glasses can be produced in some alloys.

Note: metallic glasses are not transparent. Nonetheless, the lack of a long range structure in metallic glasses does produce some unusual properties and these are employed in specialist applications. For example, Fe-Si-B metallic-glass alloys are used as magnetically soft-iron for high performance transformer cores, since the lack of long range structure prevents the processes that lead to the production of a permanent "hard" magnet.

Perhaps the most important current application of amorphous metals is in read-write compact discs (CD-RW). All compact discs (CDs) depend on changing the extent to which the disc reflects light from a laser to store information (e.g. music or computer data). Conventional CDs control using small pits that are produced mechanically in a factory and so standard CDs are read-only. In contrast, recordable CDs (CD-R) discs contain a dye layer within the disc. This dye undergoes a color change when heated. CD-R discs store information by using a laser to heat a small region of the CD, producing a localized color change that, in turn, modifies the reflectivity of the disc. Hence the term "burning a CD". CD-R discs are very cheap to produce, but suffer from the disadvantage that the color change is permanent, so that the disc is write-once-read-many (WORM). Thus, a single mistake when burning the CD and the result is only fit for use as a coaster. CD-RW discs work in a different fashion. Within the disc is a metallic layer. When this is heated locally by a laser, at a fairly high power, a small region of the metallic layer melts. The rest of the disc makes an efficient heat sink and so the small molten region cools extremely quickly and is unable to crystallize. The resulting amorphous region reflects light differently from its crystalline surroundings and so can do the same job as the pits in a conventional CD or the dye in a CD-R. The key difference is that, if the amorphous region is reheated, but with a lower laser power than before, the amorphous region becomes hot enough to allow crystallization to take place, but not so hot that the region re-melts. This erases the original information, making the CD-RW disc re-writable.

If one visits a store selling fancy glassware, one will see the phrase "crystal glassware". This is a contradiction in terms, as any given phase is either crystalline or glassy. Thus a material can be crystalline (most metals), glassy (many polymers), or a mixture of separate crystalline and glassy phases (some polymers and the class of materials called "glass ceramics").

The structure of an amorphous material is more like that of a very ***viscous*** liquid than a solid. Indeed, if one visits a very old building (say a stately home in Europe) one can see clearly that, over the course of hundreds of years, window glass will flow at room temperature!

New Words and Expressions

gemstone ['dʒemstəʊn] *n.* 宝石
polygonal [pə'lɪgənl] *adj.* 多角形的，多边形的
interface ['ɪntəfeɪs] *n.* 界面
interfacial energy [intə'feiʃəl 'enədʒi] 界面能
spherical grains 球形颗粒
viscous ['vɪskəs] *adj.* 黏的，黏性的，黏滞的

Lesson 6 Physical and Mechanical Properties of Metals

Mechanical Properties

Strength

The strength of a metal is its ability to resist changing its shape or size when external forces are applied. The four basic types of stresses are shown in Figure 2.3.

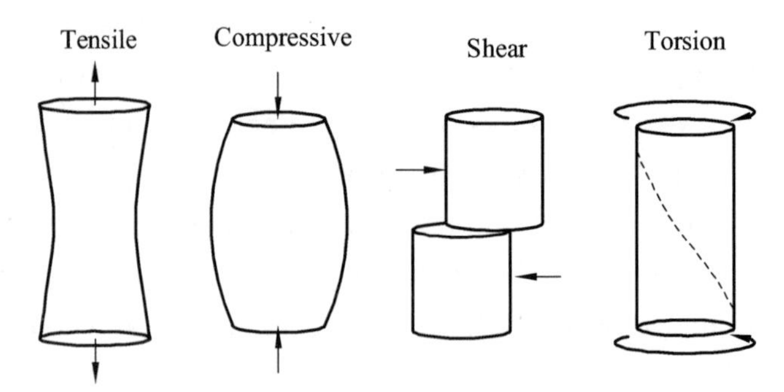

Figure 2.3 The four types of stresses

The tensile test is used to obtain information about the mechanical properties of a material. These include ductility, tensile strength, ***proportional limit***, ***elastic limit***, modulus of elasticity, ***resilience***, ***yield point***, yield strength, ***ultimate strength***, and breaking strength.

Impact strength is a term used to describe the resistance of a material to fracture resulting from impact loading. Nearly all of the metallurgical properties play a role in determining impact strength. A metal may exhibit high ductility and softness when tensile or hardness tested, but often will fracture in a brittle manner upon impact loading.

Fracture toughness is a generic term for measures of resistance to extension of a crack (i.e. , crack propagation or growth). This term is sometimes used to describe the results of an impact test, such as a test specimen exhibiting high test results in an impact test is said to have a high fracture toughness.

Elasticity

The ability of a material to strain under load and then return to its original size and shape when unloaded is called elasticity. The elastic limit (proportional limit) is the greatest load a material can withstand and still spring back to its original shape when the load is removed. Materials are subject to ***viscoelastic*** lag which means the closer a material is loaded to the elastic limit, the longer it will take for the material to return to its original size and shape upon load removal.

Yield Point

Yielding or plastic flow occurs in materials when the elastic limit has been exceeded.

Plasticity

Metals undergo plastic flow when stressed at or beyond their elastic limits. For this reason, the area of ***the stress-strain curve*** beyond the elastic limit is called the plastic range. It is this property that makes metals so useful. When enough force is applied by rolling, pressing, or hammer blows, metals can be formed when hot or cold into useful shapes.

Brittleness

A material that will not deform plastically under load is said to be brittle. Excessive cold working causes ***brittleness*** and loss of ductility. Cast iron does not deform plastically under a breaking load and is therefore brittle.

Stiffness

Stiffness is expressed by the modulus of elasticity, also called ***Young's modulus***. Within the elastic range, if the stress is divided by the corresponding strain at any given point, the result will be the modulus of elasticity for that material. Therefore, the modulus of elasticity for that material is represented by the slope of the stress-strain curve below the elastic limit.

Ductility

The property that allows a metal to deform permanently when loaded in tension is called ductility. Any metal that can be drawn into a wire is ductile. Iron, aluminum, gold, silver, and nickel are examples of ductile metals.

Ductility is measured by tensile test. Tensile specimens are measured for area and length between gage marks before and after they are pulled. The percent of elongation (increase in length) and the percent of reduction in area (decrease of area at the narrowest point) are measures of ductility. The amount of elongation before the specimen breaks is an indication of the amount of ***plastic deformation*** (clod work) that can occur in that sample of metal. These data are useful for cold forming operations in manufacturing.

Malleability

They ability of a metal to deform permanently when loaded in compression is called malleability. Metals that can be hammered or rolled into sheets are malleable. Most ductile metals are also malleable, but some very malleable metals such as lead are not very ductile and cannot be drawn into wire easily. Metals with low ductility, such as lead, can be extruded or pushed out of a die to form wire and other shapes. Some highly malleable pure metals are lead, tin, gold, silver, iron, and copper. Alloys usually tend to be less malleable and therefore require a softening heat treatment such as annealing to remove work hardening effects.

Fatigue

When metal parts are subjected to repeated loading and unloading, they may fail at stresses far below their yield strength with no sign of plastic deformation. This is called a ***fatigue failure***. When designing machine parts that are subject to vibration or cyclic loads, fatigue strength may be more important than ultimate tensile or yield strength.

The fatigue limit, or endurance limit, is the maximum load in pounds per square inch that can be applied an infinite number of times without causing failure. Ten million loading cycles usually considered enough to establish fatigue limits.

Creep Strength

Creep is a continuing, slow plastic flow at a stress below the elastic limit of a metal. Creep is usually associated with high temperatures but does occur to some extent at normal temperatures. Low temperature creep can take months or years to alter machine parts that are habitually left in a stressed condition. As the temperature increases, creep becomes more of a problem. Creep strength for a metal is given in terms of an allowable amount of plastic flow (creep) per 1000-hour period.

Metals at Low Temperatures

As the temperature decreases, the strength, hardness, and modulus of elasticity increase for almost all metals. The effect of temperature drop on ductility separates metals into two groups: those that become brittle at low temperatures and those that remain ductile.

Metals of the group that remain ductile show a slow, steady decrease in ductility with a drop in temperature. Metals in the group that become brittle at low temperatures show a temperature range where ductility and, most important, toughness drop rapidly. This rang is called the ***transition zone***.

Physical Properties

Conductivity of Metals

Metals conduct heat better than nonmetals. Silver conducts heat the best of all metals. This ability to conduct heat and ability to conduct electricity are related. Since silver is the best heat conductor, it is also the best electrical conductor. Note that in all cases the pure metals are better conductors than their alloys. Pure copper would be a better choice for electrical wiring than a copper alloy. A pure aluminum automobile radiator would conduct heat away from the water inside better than an aluminum alloy radiator would.

Thermal Expansion

In almost all cases, solids become larger when heated and smaller when cooled. Each substance expands and contracts at a different rate. This rate is expressed in inches per inch degree Fahrenheit and is called the ***coefficient of thermal expansion***.

New Words and Expressions

proportional limit [prə'pɔːʃənəl 'limit] 比例极限
elastic limit [i'læstik 'limit] 弹性极限
resilience [rɪ'zɪliəns] *n.* 弹性，弹力，快速恢复的能力，回弹
Yield point [jiːld pɔint] 屈服点
ultimate strength ['ʌltimit streŋθ] 极限强度
impact strength 冲击强度
fracture toughness ['fræktʃə 'tʌfnis] 断裂韧性
crack propagation [kræk ˌprɔpə'geɪʃən] 裂纹扩展
viscoelastic [ˌvɪskəʊɪ'læstɪk] *adj.* 黏弹性的
stress-strain curve [strɛs stren kɚv] 应力应变曲线
brittleness ['brɪtlnəs] *n.* 脆性，脆度
Young's modulus 杨氏模量
plastic deformation ['plɑːstik ˌdiːfɔː'meɪʃən] 塑性变形
fatigue failure [fə'tiːg 'feiljə] 疲劳破坏
transition zone [træn'ziʃən zəun] 转变区
Fahrenheit ['færənhaɪt] *adj.* 华氏温度计的；*n.* 华氏温度计
coefficient of thermal expansion 热膨胀系数

Reading Material

Corrosion

Ferrous metals have a tendency to ***rust***, because the iron in them reacts with oxygen in the environment to form iron oxide, or rust. This process is called corrosion. The process of corrosion can involve the metal coming into contact with water, which contains oxygen or another element that is strongly negative. Two different metals in contact with an ***electrolyte***, typically water, compose an electrolytic cell, or ***galvanic cell***. One metal acts as the positive electrode, or ***anode***, and the other acts as the negative electrode, or ***cathode***. This process is called ***galvanic corrosion***. Galvanic corrosion is an ***electrochemica***l process, which ***erodes*** the anode. The oxide or iron forms a larger crystal than the steel itself, causing the rust to buckle away from the surface of the metal and to flake off, eroding the surface. If allowed to continue, this flaking will eventually eat through the metal.

For example, take two metals, aluminum and steel, joined together. Look at the relative positions of the two metals in the galvanic series. Aluminum is higher than steel. Therefore, aluminum is the more positive, or anodic. Aluminum will give up electrons to the conducting liquid and will corrode. In some situations, this fact is used as an advantage. For instance, brass and bronze contain copper. The copper will react with chlorine, oxygen, or sulfur to form a green or black coating of copper chloride, copper oxide, or copper sulfate, respectively. This coatings stick to the surface of the metal and protect it from further attack. This can be observed in statues where the artist artificially ***corrodes*** the metal for visual appeal and to protect the base metal.

There are several factors that may speed the corrosion process. Among these factors are an increase in temperature. The presence of certain gases, environmental factors such as acid rain, metal fatigue, cold working/forming, and other similar factors. Alloying elements tend to retard the corrosion process by protecting the metal against oxide formation. In addition, various coatings are used to protect the surface of the metal.

Corrosion also occurs in other materials, such as plastics and elastomers. Corrosion in these materials tend to deteriorate them, reducing their mechanical properties and making them more brittle. Corrosion in old tires is apparent as weathering and rot, generally caused by environmental factors. Other materials such as glass and ceramics are oxides, sulfides, and other natural compounds that resist deterioration.

Corrosion is often an electrochemical process. Therefore, poor conductors such as ceramics are corrosion-resistant. The process requires an anode, a cathode, and some form of electrolyte. If you were to immerse two metals, such as aluminum and steel, in water, galvanic corrosion will occur. One metal acts as an anode and the other as the cathode. Oxidation or the loss of electrons takes place at the anode, while reduction or the gaining of electrons takes place at the cathode. The water serves as the conductor or electrolyte. As mentioned earlier, the corrosion process is electrochemical. A potential and a path for current flow must exist. The anode and the cathode can exist in the same material. For example, two alloying elements in the same material within a humid environment may produce the necessary conditions for corrosion.

There are several ways to protect against corrosion. Corrosion can be prevented or lessened by coatings, design considerations, environmental control, and alloying, among others.

New Words and Expressions

rust [rʌst] *n.* 铁锈；*vt.* 使生锈

electrolyte [ɪ'lektrəlaɪt] *n.* 电解质，电解液

galvanic cell [gæl'vænik sel] 原电池

anode ['ænəʊd] *n.* 阳极，正极

cathode ['kæθəʊd] *n.* 阴极，负极

electrochemical [ɪˌlektrəʊ'kemɪkəl] *adj.* 电化学的

erode [ɪ'rəʊd] *v.* 侵蚀，腐蚀

corrode [kə'rəʊd] *v.* 使腐蚀，侵蚀

UNIT 3 CERAMICS

Lesson 7 Introduction to Ceramics

Definition of Ceramics

The word ceramic, derives its name from the Greek *keramos*, meaning "pottery", which in turn is derived from an older Sanskrit root, meaning "to burn". The Greeks used the term to mean "burnt stuff" or "burned earth". Thus the word was used to refer to a product obtained through the action of fire upon earthy materials.

Ceramics make up one of three large classes of solid materials. The other material classes include metals and polymers. The combination of two or more of these materials together to produce a new material whose properties would not be attainable by conventional means is called a ***composite***.

Ceramics can be defined as solid compounds that are formed by the application of heat, and sometimes heat and pressure, comprising at least two elements provided one of them is a non-metal or a nonmetallic elemental solid. The other element(s) may be a metal(s) or another nonmetallic elemental solid(s). A somewhat simpler definition was given by Kingery who defined ceramics as, "the art and science of making and using solid articles, which have, as their essential component, and are composed in large part of ***inorganic nonmetallic materials***". In other words, what is neither a metal, a semiconductor or a polymer is a ceramic.

To illustrate, consider the following examples: Magnesia, or MgO, is a ceramic since it is a solid compound of a metal bonded to the nonmetal O_2. Silica is also a ceramic since it combines an nonmetallic elemental solids (NMES) and a nonmetal. Similarly, TiC and ZrB_2 are ceramics since they combine metals (Ti, Zr) and the NMES (C, B)- SiC is a ceramic because it combines two NMESs. Also note ceramics are not limited to ***binary compounds***: $BaTiO_3$, $YBa_2Cu_3O_3$, and Ti_3SiC_2 are all perfectly respectable class members.

It follows that the oxides, nitrides, borides, carbides, and silicides (not to be confused with silicates) of all metals and NMESs are ceramics; which, needless to say, leads to a vast number of compounds. This number becomes even more daunting when it is appreciated that the silicates are also, by definition, ceramics. Because of the abundance of oxygen and silicon in nature, silicates are ubiquitous; rocks, dust, clay, mud, mountains, sand—in short, the vast majority of the earth's crust—are composed of silicate based minerals. When it is also appreciated that even cement, bricks, and concrete are essentially silicates, the case could be made that we live in a ceramic world.

Archeologists have uncovered human-made ceramics that date back to at least 24 000 BC. These ceramics were found in what was formerly Czechoslovakia and were in the form of animal and human figurines, slabs, and balls. These ceramics were made of animal fat and bone mixed with bone ash and a fine claylike material. After forming, the ceramics were fired at temperatures between 500-800 °C in domed and horseshoe shaped kilns partially dug into the ground with loess walls. While it is not clear what these ceramics were used for, it is not thought to have been a utilitarian one. The first use of functional pottery vessels is thought to be in 9000 BC. These vessels were most likely used to hold and store grain and other foods.

Classification of Ceramics

Most people associate the word ceramics with pottery, sculpture, sanitary ware, tiles, etc. And whereas this view is not incorrect, it is incomplete because it considers only the traditional, or ***silicate-based ceramics***. Today the field of ceramic science or engineering encompasses much more than silicates and can be divided into traditional and advanced ceramics. Before the distinction is made, however, it is worthwhile to trace the history of ceramics and people's association with them.

It has long been appreciated by our ancestors that some muds, when wet, were easily moldable into shapes that upon heating became rigid. The formation of useful articles from fired mud must constitute one of the oldest and more fascinating of human endeavors. Fired-clay articles have been traced to the dawn of civilization. The usefulness of these new materials, however, was limited by the fact that when fired, they were ***porous*** and thus could not be used to carry liquids. Later the serendipitous discovery was made that when heated and slowly cooled, some sands tended to form a transparent, water-impervious solid, known today as glass. From that point on, it was simply a matter of time before ***glazes*** were developed that rendered clay objects not only watertight, but also quite beautiful.

With the advent of the industrial revolution, structural clay products, such as bricks and heat-resistant refractory materials for the large-scale smelting of metals were developed. And with the discovery of electricity and the need to distribute it, a market was developed for electrically insulating silicate-based ceramics.

Traditional ceramics are characterized by mostly silicate-based porous microstructures that are quite coarse, nonuniform, and multiphase. They are typically formed by mixing clays and ***feldspars***, followed by forming either by slip casting or on a potter's wheel, firing in a flame kiln to sinter them, and finally glazing.

In a much later stage of development, other ceramics that were not clay or silicate-based depended on much more sophisticated raw materials, such as binary oxides, carbides, ***perovskites***, and even completely synthetic materials for which there are no natural equivalents. The microstructures of these modern ceramics were at least an order of magnitude finer and more homogeneous and much less porous than those of their traditional counterparts. It is the latter — the advanced or technical ceramics.

Most of these advanced ceramics were development within the past 60 years, which is quite recent in materials technology. Much of that development resulted from aerospace research and development (R&D), but gradually, the costs associated with advanced ceramics have made them feasible for more earthly applications, such as in the automobile, sports, and machine-tool industries. Most are actually synthetic ceramics that are produced from fine, relatively poor powders using new technology that includes microwaves, electron beams, and polymer chemistry. Figure 3.1 lists some differences between traditional and advanced ceramics, including some applications of each category of ceramic.

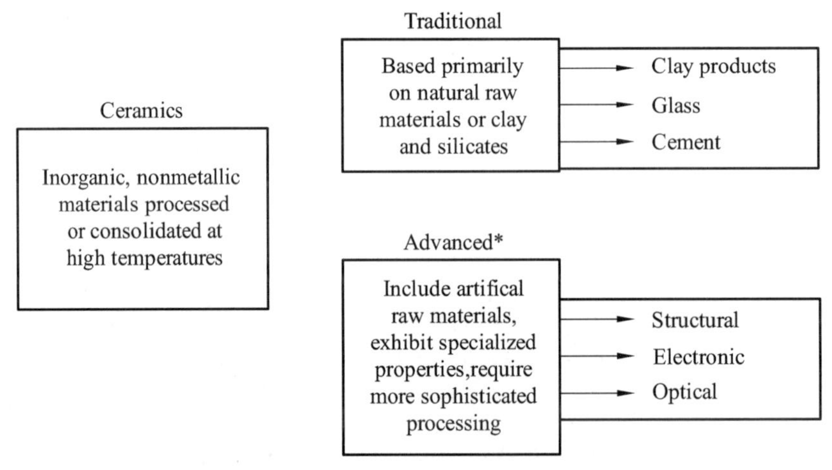

Figure 3.1 Broad classifications of ceramics

*Fine Ceramics, or Engineered ceramics, or New ceramics, or Valve-added ceramics

Advanced ceramics are receiving attention for wide-ranging scitech development because they offer the designer properties such as light weight, good strength at elevated temperatures, and wear resistance. Many of these newer ceramics are composites reinforced with whiskers and fibers to improve their fracture toughness, so they do not break catastrophically as we generally expect dishes and coffee cups to do when we drop them. The wear resistance of ceramics makes them valuable as coatings for cutting tools, surgical instruments, punches, and dies. For a modest cost, it is possible to apply a TiN coating only a few micrometers thick to tools, which can extend their life 7 to 15 times. You may have seen golden TiN coatings on twist drills in the hardware store.

In the ongoing search to produce superior materials with a minimum of energy (low temperature , low pressure, and minimum number of steps), researchers have turned to nature to replicate its processing techniques. Biomimickry or ***biomimetics*** is the attempt to mimic or replicate natural processes. For example, seashells are intricate, lamellar ceramic composites made with low energy yet at a nanometri scale, yielding complex patterns that hold promise for ceramic electronic devices and similar advanced applications.

Figure 3.2 represents a taxonomy of ceramic materials based on their applications. Note that other classifications of ceramics can be developed based on such differences as their processing. New or advanced ceramics do not possess the plasticity and ***formability*** of clay when mixed with water, which necessitates new techniques in processing to be discussed later in this module. This module should help you to gain a perspective on the structure, properties, selection, and basic processes in ceramics, both traditional and advanced. In your study of ceramics, ponder their nature and their ability to compete with metals, polymers, and composites. Be alert to news articles and advertisements that describe emerging applications of advanced ceramics. Also, imagine where there may be new opportunities for the use of ceramics. This information will serve you well in our advancing age of materials and prepare you for the dramatic changes in this era.

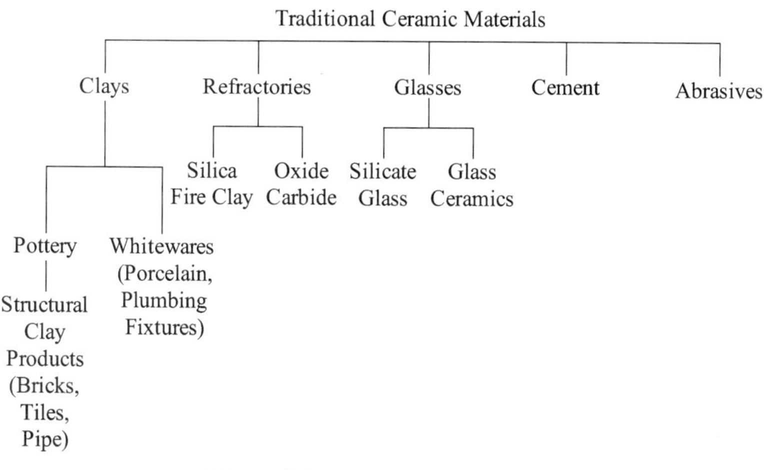

(a) Traditional ceramic materials

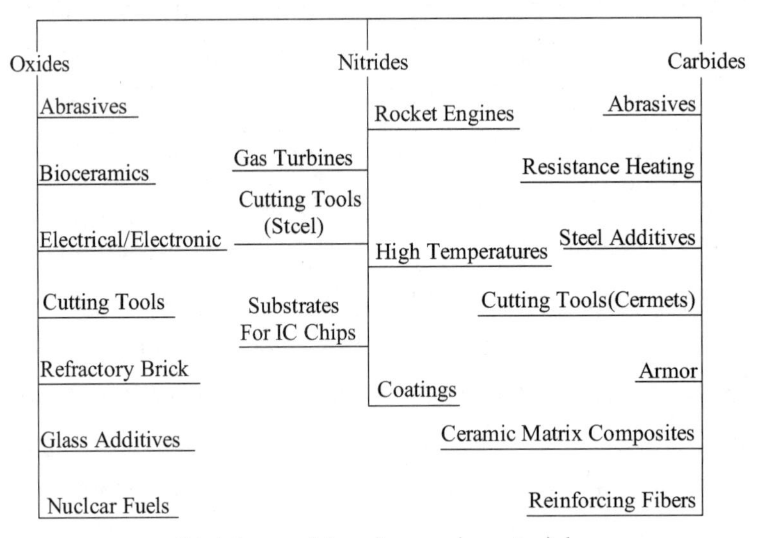

(b) Advanced (new) ceramic materials

Figure 3.2 Taxonomy of ceramic materials based on applications

New Words and Expressions

composite ['kɒmpəzɪt] *n.* 复合材料
inorganic nonmetallic materials 无机非金属材料
binary compounds 二元化合物
silicate-based ceramics 硅酸盐基陶瓷
porous ['pɔːrəs] *adj.* 多孔的
glaze [gleɪz] *v.* 上釉
feldspar ['feldspɑː(r)] *n.* 长石
perovskite [pə'rɒvzkaɪt] *n.* 钙钛矿
biomimetics [baɪɒmɪ'metɪks] *n.* 仿生学
formability [fɔːmə'bɪlɪtɪ] *n.* 可成形性

Reading Material

Reviews of Materials physics: Structural Materials

At the turn of the last century, mankind's use of structural materials was limited primarily to metals, particularly iron and its alloys, ceramics (most notably ***Portland***

cement), and polymers, which were limited to naturally occurring rubbers, ***glues***, and fibers. Composites, as a concept were nonexistent even though wood and animals, each composed of different materials, were used in a variety of ways. However, the uses of alloying to enhance the strength of ***lightweight*** materials, such as ***pewter***, or copper additions to aluminum, were established techniques, known well before this century. This knowledge was used to build the first dirigibles. The useful nature of a material was often understood through serendipity and not through an understanding of its structure or the relation between structure and properties. We still cannot predict in any quantitative way the evolution of structure with deformation or processing of a material. However, we have come a very long way from the situation that existed a hundred years ago, thanks to the contributions of twentieth century science to our understanding of atomic arrangement and its determination in a material. Our classification of materials by symmetry considerations came into existence once atomic arrangement became known. To the seven crystal systems and amorphous structures, typified by the glasses and liquids, we can now add ***quasi crystals*** and ***molecular phases***, such as ***fullerenes*** and nanotubes, in a crystalline solid.

The crystal systems define perfect crystals. At finite temperatures, the crystals are no longer perfect but contain defects. It is now understood that these defects are responsible for atomic transport in solids. In fact, the structural properties of materials are not only a function of the inherent strength of a material but also of the defects that may be present. We know that aluminum is soft because crystallographic defects, called dislocations, can be readily generated and moved in this metal. In contrast, in alumina (Al_2O_3), dislocation generation and motion are difficult; hence alumina can be strong but brittle at room temperature. The addition of copper or manganese to aluminum creates second-phase precipitates, which inhibit the motion of dislocations, thus enhancing its strength-to-weight ratio. Our ability to improve the strength-to-weight ratio in materials has increased more than tenfold during the twentieth century. This is to be compared with a change of less than ten over the last twenty centuries. Much of the increase in this century has come from an understanding of the relationship between the processing of materials and their structure. The highest strength-to-weight ratios have been achieved in materials in the form of fibers and nanotubes. In these structures, dislocations either do not exist or do not move.

Most structural materials are not single crystals. In fact, they consist of a large number of crystals joined at interfaces, which in single-phase materials are called grain boundaries. These interfaces can, for example, influence the mechanical and electrical properties of materials. At temperatures where the grain boundary diffusion rate is low, a small grain size enhances the strength of a material. However, when the grain boundary diffusion rate is high, the material can exhibit very large elongation under a tensile load (superplastic behavior), or can exhibit high creep rates under moderate or small conditions of loading. In demanding high-temperature environments, such as the engine of a modern aircraft, grain boundaries are eliminated so that a complex part, such as a turbine blade, made of a nickel alloy, is a single crystal. Thus the use of materials for structural purposes requires an understanding of the behavior of defects in solids. This is true for metallic, ceramic, and glassy materials.

Both ceramics and glasses were known to ancient civilizations. Ceramics were used extensively in pottery and art. The widespread use of ceramics for structural purposes is largely limited by their brittle behavior. This is now well understood, and schemes have been proposed to overcome brittleness by controlling the propagation of cracks. In metals, dislocations provide the microscopic mechanism that carries energy away from the tip of a crack, thereby blunting it. In ceramics, the use of phase transformations induced at the tip of a propagating crack is one analog of dislocations in metals. Other schemes involve the use of bridging elements across cracks so as to inhibit their opening and hence their propagation. Still another scheme is to use the frictional dissipation of a sliding fiber embedded in a matrix not only to dissipate the energy of crack propagation, but also, if the crack propagates through the material, to provide structural integrity. Use of these so-called fault tolerant materials requires both an understanding of mechanical properties and control over the properties of interfaces to enable some sliding between the fiber and the matrix without loss of adhesion between them. Such schemes rely either on composite materials or on microstructures that are very well controlled.

The widespread use of silicate glasses, ranging from windows to laptop displays, is only possible through the elimination of flaws, which are introduced, for example, by inhomogeneous cooling. These flaws, which are minute cracks, are eliminated during processing by controlling the cooling conditions, as in a tempered glass, and also by introducing compressive strains through composition modulations.

There are a number of fibers that are available for use with ceramics, polymers, and metals to form composite materials with specific applications; these include carbon fibers, well known for their use in golf clubs and fishing rods, and silicon carbide or nitride fibers. Optical fibers, which are replacing copper wires in communication technologies, owe their widespread use not only to their optical transparency, but also to improvements in their structural properties. Fibers must withstand mechanical strains introduced during their installation and operation.

The use of composite materials in today's civilization is quite widespread, and we expect it to continue as new applications and "smart" materials are developed. An outstanding example of a functional composite product comes from the electronics industry. This is a substrate, called a package, which carries electronic devices. Substrates are complicated three-dimensionally designed structures, consisting of ceramics, polymers, metals, semiconductors, and insulators. These packages must satisfy not only structural needs but also electrical requirements.

Although we have made great progress over the last hundred years in materials physics, our microscopic understanding of the physics of deformation (particularly in noncrystalline solids), fracture, wear, and the role of internal interfaces is still far from complete. There has been considerable progress in computer simulation of some of these issues. For example, there is now a concerted effort to model the motion of dislocations, during deformation, in simulations of simple metallic systems. We anticipate that within the next decade, as computational power continues to increase, many of these problems will become tractable. The ultimate goal is to design a structural component for a set of specified environmental conditions and for a predictable lifetime.

New Words and Expressions

Portland cement *n.* 波特兰水泥，普通水泥，硅酸盐水泥
glue [gluː] *n.* 胶，胶水
lightweight ['laɪtweɪt] *adj.* 轻质的
pewter ['pjuːtə(r)] *n.* 白蜡，白镴
quasi crystals 准晶体
molecular phases 分子相
fullerenes ['fʊləriːnz] *n.* 富勒烯

Lesson 8 Structure of Ceramics

Most crystalline inorganic compounds are based on nearly close-packing of the anions (generically referred to as O or X, though oxygen is the most common anion) with metal atom cations (generically called M or A) placed interstitially within the anion lattice. A summary of some simple ionic structures and their corresponding coordination numbers is given in Table 3.1. Some of the most common ceramic and ionic crystal structures are described in the following sections.

Table 3.1 Table of some simple ionic structures and their corresponding *coordination numbers*

Anion Packing	Coordination Number of M and O	Sites by Cations	Structure Name	Examples
Cubic close-packed	6 : 6 MO	All oct.	Rock salt	NaCl, KCl, LiF, KBr, MgO, CaO, SrO, BaO, CdO, VO, MnO, FeO, CoO, NiO
Cubic close-packed	4 : 4 MO	1/2 tet.	Zinc blende	ZnS, BeO, SiC
Cubic close-packed	4 : 8 M_2O	All tet.	Antifluorite	Li_2O, Na_2O, K_2O, Rb_2O, sulfides
Distorted cubic close-packed	6 : 3 MO_2	1/2 oct.	Rutile	TiO_2, GeO_2, SnO_2, PbO_2, VO_2, NbO_2, TeO_2, MnO_2, RuO_2, OsO_2, IrO_2
Cubic close-packed	12 : 6 : 6 ABO_3	1/4 oct. (B)	Perovskite	$CoTiO_3$, $SrTiO_3$, $SrSnO_3$, $SrZrO_3$, $SrHfO_3$, $BaTiO_3$
Cubic close-packed	4 : 6 : 4 AB_2O_4	1/8 tet. (A) 1/2 oct. (B)	Spinel	$FeAl_2O_4$, $ZnAl_2O_4$, $MgAl_2O_4$
Cubic close-packed	4 : 6 : 4 B $(AB)O_4$	1/8 tet. (B) 1/2 oct. (A, B)	Spinel (inverse)	$FeMgFeO_4$, $MgTiMgO_4$
Hexagonal close-packed	4 : 4 MO	1/2 tet.	Wurtzite	ZnS, ZnO, SiC
Hexagonal close-packed	6 : 6 MO	All oct.	Nickel arsenide	NiAs, FeS, FeSe, CoSe
Hexagonal close-packed	6 : 4 M_2O_3	2/3 oct.	Corundum	Al_2O_3, Fe_2O_3, Cr_2O_3, Ti_2O_3, V_2O_3, Ga_2O_3, Rh_2O_3
Hexagonal close-packed	6 : 6 : 4 ABO_3	2/3 oct. (A, B)	Ilmenite	$FeTiO_3$, $NiTiO_3$, $CoTiO_3$
Hexagonal close-packed	6 : 4 : 4 A_2BO_4	1/2 oct. (A) 1/8 tet. (B)	Olivine	Mg_2SiO_4, Fe_2SiO_4
Simple cubic	8 : 8 MO	All cubic	CsCl	CsCl, CsBr, CsI
Simple cubic	8 : 4 MO_2	1/2 cubic	Fluorite	ThO_2, CeO_2, PrO_2, UO_2, ZrO_2, HfO_2, NpO_2, PuO_2, AmO_2
Connected tetrahedra	4 : 2 MO_2	—	Silica types	SiO_2, GeO_2

Rock Salt Structure

The compounds NaCl, MgO, MnS, LiF, FeO, and many other oxides such as NiO and CaO have the so-called ***rock salt structure***, in which the anions, such as Cl or O, are arranged in an FCC array, with the cations placed in the ***octahedral*** interstitial sites, also creating an FCC array of cations, as shown in Figure 3.3. The primitive cell of the rock salt structure, however, is simple cubic, with anions at four corners, and cations at alternating corners of the cube, as given by one of the ***quadrants*** in Figure 3.3. As indicated in Table 3.1, the coordination number of both anions and cations is 6, so the cation-anion radius ratio is between 1.37 and 2.42.

Diamond Structure

Another common ceramic structure arises when the ***tetrahedral*** sites in an FCC array of anions are occupied. For example, an FCC array of sulfur anions with Zn ions in the tetrahedral positions gives the compound ZnS and results in the ***zinc blende structure*** (see Figure 3.4). The atoms can be all alike also, for example carbon, in which case diamond results. In either case, the atoms in the tetrahedral sites have a coordination number of four (by definition). Many important compounds have the zinc blende or diamond structure, including SiC.

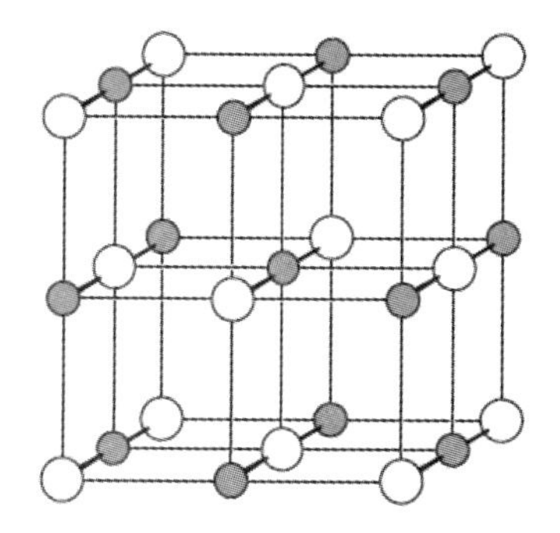

Figure 3.3 The rock salt crystal structure

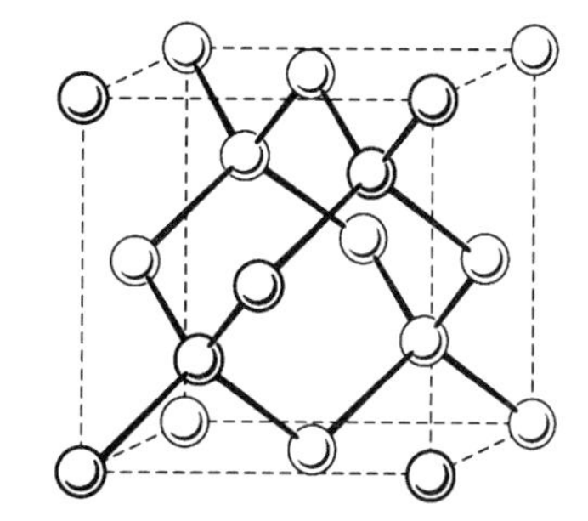

Figure 3.4 The diamond (zinc blende) crystal structure

Spinel Structure

Many compounds are formed when there is more than one metal cation in the lattice. Such is the case with the ***spinel*** structure, which has the general formula AB_2O_4, where A and B are different metal cations, such as in ***magnesium aluminate***, $MgAl_2O_4$ (see Figure 3.5). This structure can be viewed as a combination of the rock salt and zinc blende structures. The anions, usually oxygen, are again placed in an FCC array. In a normal spinel, the divalent A ions are on tetrahedral sites and the trivalent B atoms are on octahedral sites. In an inverse spinel, divalent A atoms and half of the trivalent B atoms are on octahedral sites, with the other half of the B^{3+} atoms on tetrahedral sites. Many of the ferrites, such as Fe_3O_4 (in which iron has two different coordination states), have the inverse spinel structure.

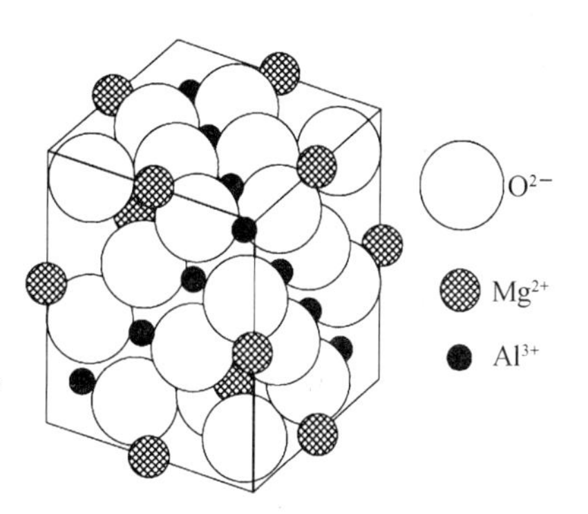

Figure 3.5 The spinel crystal structure of $MgAl_2O_4$

Other Important Ceramic Structures

There are many additional crystal structures that arise due to compounds of both simple and complex ***stoichiometry***. These structures generally have specific names associated with them that have developed out of geology and crystallography over many years. For example, the ***corundum*** structure is common to Al_2O_3, the ***rutile*** structure comes from one of the forms of TiO_2, and a number of important ceramics, such as CaF_2, have the ***fluorite*** structure. One structure with current technological importance is the perovskite structure (see the $CaTiO_3$ perovskite structure in Figure 3.6). Many important ceramics with unique electrical and dielectric properties have the perovskite structure, including barium titanate, $BaTiO_3$, and ***high-temperature superconductors*** (HTS). The perovskites have the general formula of ABO_3, but in the case of most superconductors, the A cation consists of more than one type of atom, such as in $Y_1Ba_2Cu_3O_{7-x}$, or the so-called "1-2-3" superconductor, in which the perovskite structure is tripled, and one yttrium atom is replaced for every third barium atom; there is usually less than the stoichiometric nine oxygen atoms required in this structure in order for enough oxygen vacancies to form and superconductivity to result.

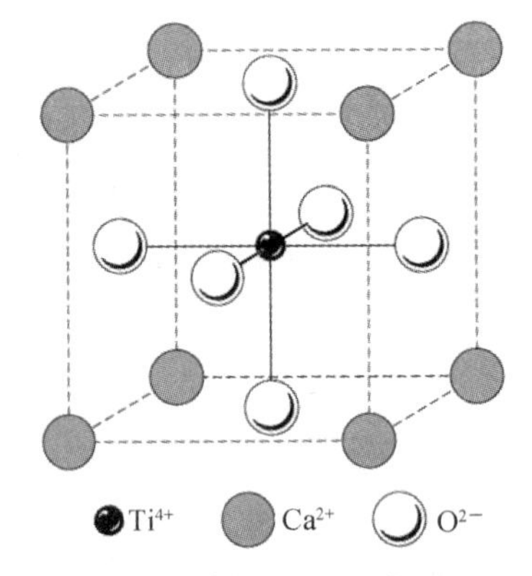

Figure 3.6 The perovskite crystal structure of $CaTiO_3$

New Words and Expressions

coordination number [kəuˌɔːdn'eɪʃən 'nʌmbə] 配位数

rock salt structure *n.* 岩盐结构

octahedral [ˌɒktə'hiːdrəl] *adj.* 八面体的

quadrant ['kwɒdrənt] *n.* 象限

tetrahedral ['tetrə'hedrəl] *adj.* 四面体的

zinc blende structure [ziŋk blend 'strʌktʃə] 闪锌矿（型）结构

spinel [spɪ'nel] *n.* 尖晶石

magnesium aluminate [mæg'niːziːəm ə'ljuːmineit] 铝酸镁
stoichiometry [ˌstɔɪkɪ'ɒmɪtrɪ] *n.* 化学计算（法），化学计量学
corundum [kə'rʌndəm] *n.* 刚玉，金刚砂
rutile ['ruːtiːl] *n.* 金红石
fluorite ['flʊəraɪt] *n.* 萤石，氟石
high-temperature superconductors 高温超导体

Reading Material

Silicate Structures

The silicates, made up of base units of silicon and oxygen, are an important class of ceramic compounds that can take on many structures, including some of those we have already described. They are complex structures that can contain several additional toms such as Mg, Na, K. What makes the silicates so important is that they can e either crystalline or amorphous (glassy) and provide an excellent opportunity to compare these two ***disparate*** types of structure. Let us first examine the crystalline state, which will lead us into the amorphous state. The structural unit for the simplest silicate, SiO_2, also known as silica, is the tetrahedron see Figure 3.7). This is the result of applying Pauling's principles to a compound between silicon and oxygen.

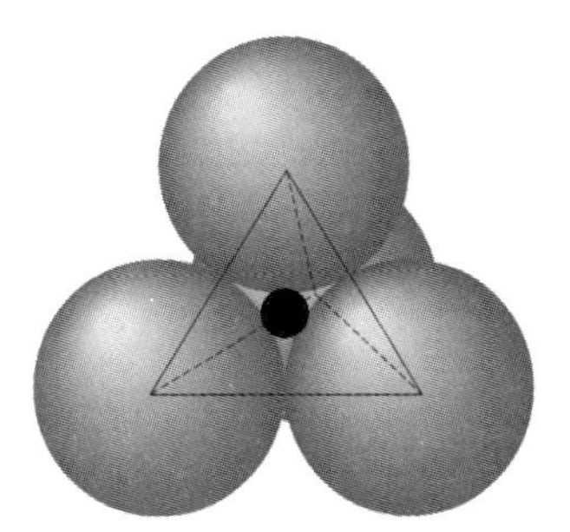

Figure 3.7 The $(SiO_4)^{4-}$ tetrahedron

The data in Table 3.2 indicate that the anion/cation ratio in SiO_2 is R_O/R_{Si} = 1.32/0.39 = 3.3, which dictates the tetrahedron as the base structural unit. Note that the SiO_4 tetrahedron has a formal charge of −4, which must be neutralized with cations, such as other Si atoms, in real compounds. Pauling's second rule tells us that the bond strength in silicon is 1, and the third and fourth rules tell us that corners of the tetrahedra are generally shared. This is not always the case, and different macroscopic silicate structures result depending on how the tetrahedra are combined. Corners, edges, or faces of tetrahedra can be shared. As the nature of combination of the tetrahedra changes, so must the O/Si ratio, and charge neutrality is maintained through the addition of cations. These structures are summarized in Table 3.2 and will be described separately.

Table 3.2 Structural units observed in crystalline silicates

O/Si Ratio	Silicon-oxygen Groups	Structural Units	Examples
2	SiO_2	Three-dimensional network	Quartz
2.5	Si_4O_{10}	Sheets	Talc
2.75	Si_4O_{11}	Chains	Amphiboles
3.0	SiO_3	Chains, rings	Pyroxenes, beryl
3.5	Si_2O_7	Tetrahedra sharing one oxygen ion	Pyrosilicates
4.0	SiO_4	Isolated orthosilicate tetrahedra	Orthosilicates

Crystalline Silicate Network

When all four corners of the SiO_4 tetrahedra are shared, a highly ordered array of networked tetrahedra results, as shown in Figure 3.8. This is the structure of ***quartz***, one of the crystalline forms of SiO_2. Notice that even though the O/Si ratio is exactly 2.0, the structure is still composed of isolated $(SiO_4)^{4-}$ tetrahedra. Each oxygen on a corner is shared with one other tetrahedron, however, so there are in reality only two full oxygen atoms per tetrahedron. There are actually several structures, or ***polymorphs***, of crystalline silica, depending on the temperature. Quartz, with a density of 2.655 $g \cdot cm^{-3}$, is stable up to about 870 °C, at which point it transforms into tridymite, with a density of 2.27 $g \cdot cm^{-3}$. At 1470 °C, ***tridymite*** transforms to ***cristobalite*** (density = 2.30 $g \cdot cm^{-3}$), which melts at around 1710 °C. There are "high" and "low" forms of each of these structures, which result from slight, albeit rapid, rotation of the silicon tetrahedra relative to one another.

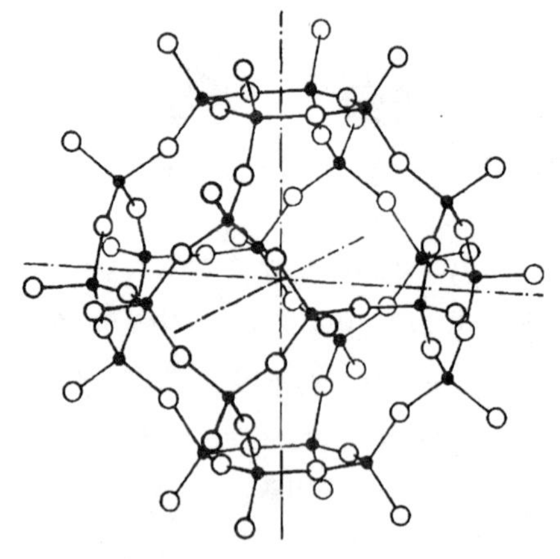

Figure 3.8 The structure of quartz, showing the three-dimensional network of SiO_4 tetrahedra

Silicate Sheets

If three of the four corners of the $(SiO_4)^{4-}$ tetrahedron are shared, repeat units of $(Si_2O_5)^{2-}$ or $(Si_4O_{10})^{4-}$ result, with a corresponding O/Si of 2.5. Table 3.2 tells us, and Figure 3.9 shows us, that sheet structures are the result of sharing three corners. In these structures, additional cations or network modifiers, such as Al^{3+}, K^+, and Na^+, preserve charge neutrality. Through simple substitution of selected silicon atoms with aluminum atoms, and some hydroxide ions (OH^-) for oxygen atoms, complex and amazing sheet structures can result. One such common example is ***muscovite***, $K_2Al_4(Si_6Al_2)O_{20}(OH)_4$, more commonly known as ***mica*** (Figure 3.10). The large potassium ions between layers create planes that are easily cleaved, leading to the well-known thin sheets of mica that can be made thinner and thinner in a seemingly endless fashion. It is, in fact, possible to obtain atomically smooth surfaces of mica.

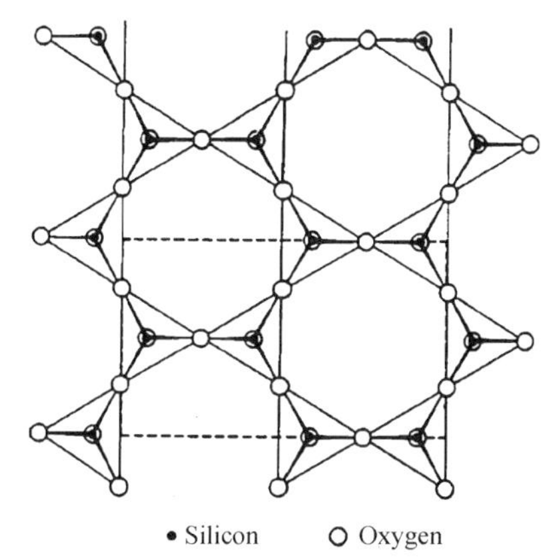

Figure 3.9 Top view of a silicate sheet structure resulting from sharing three corners of the SiO_4 tetrahedra

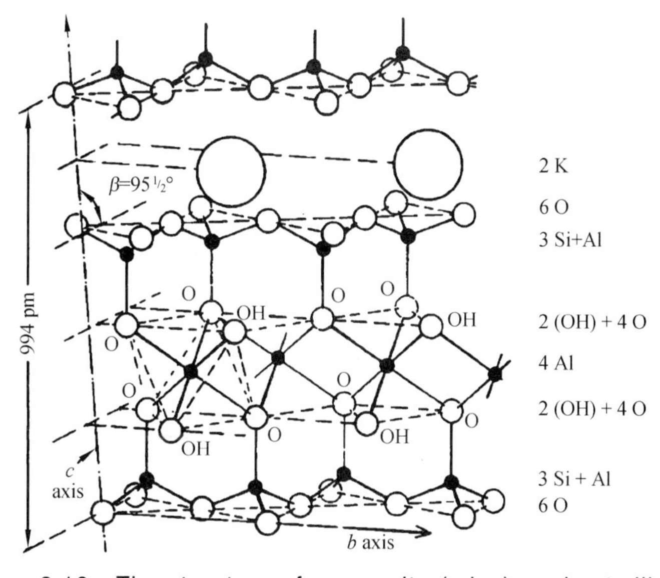

Figure 3.10 The structure of muscovite (mica), a sheet silicate

Silicate Chains and Rings

Sharing two out of the four corners of the SiO_4 tetrahedra results in chains. The angle formed between adjacent tetrahedra can vary widely, resulting in unique structures such as rings (see Figure 3.11). In all cases, when only two corners are shared, the repeat unit is $(SiO_3)^{2-}$, and the O/Si ratio is 3.0.

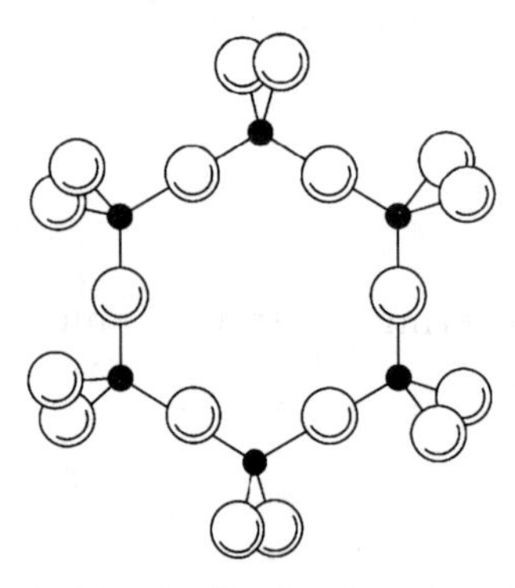

Figure 3.11 A silicate ring, beryl, with two corners of the SiO_4 tetrahedra shared

Slight variations in the O/Si ratio can also take place, and result in partially networked structures such as double chains, in which two silicate chains are connected periodically by a bridging oxygen. Asbestos is such a double chain, with O/Si ratio = 2.75.

Pyrosilicates

One corner of the SiO_4 tetrahedron shared results in a $(Si_2O_7)^{6-}$ repeat unit and a class of compounds called the ***pyrosilicates***. Again, counterions are necessary to maintain charge neutrality. The pyrosilicates are nonnetworked and have an O/Si ratio of 3.5.

Orthosilicates

Finally, no tetrahedral corners shared gives an O/Si ratio of 4.0, and it results in isolated $(SiO_4)^{4-}$ tetrahedra. These class of materials are referred to as the ***orthosilicates***.

New Words and Expressions

disparate ['dɪspərət] *adj.* 全异的，完全不同的

quartz [kwɔːts] *n.* 石英

polymorphs *n.* 多晶形，多形体，同质多形体

tridymite ['traɪdəmaɪt] *n.* 磷石英

cristobalite [krɪs'təʊbəlaɪt] *n.* 方石英

muscovite *n.* 白云母

mica ['maɪkə] *n.* 云母

pyrosilicate *n.* 焦硅酸盐

orthosilicate [ɔːθə'sɪləkeɪt] *n.* 正硅酸盐

Lesson 9 Properties of Ceramics (Ⅰ)

We have shown how the atomic structure and microstructure of ceramics differ from metals and plastics; now we shall discuss some of the property differences that set ceramics apart from other engineering materials. Table 3.3 compares ceramics with other materials by listing some of their properties. We will discuss these to develop a general feeling for the properties of ceramics.

Table 3.3 Property comparing of engineering ceramics to other engineering materials

Property	Ceramic	Metals	Plastics
Mechanical			
Tensile strength	100 max.	300 max.	30 max.
Yield strength	Does not yield	250 max.	25 max.
Elongation	Zero	Up to 50%	Can be 100%
Compressive strength	600 max.	400 max.	40 max.
Impact	Poor	Poor to excellent	Fair
Hardness	Can be 2600	900 max.	100 max.
Tensile modulus	Can be 90×10^6	Can be 40×10^6	Can be 1×10^6
Creep strength	Excellent	Good	Poor
Fatigue strength	OK in compression	Good to excellent	Poor
Physical			
Density	0.1 to 0.6	0.06 to 0.8	0.03 to 0.1
Melting point	>5000 °C	Up to 2760 °C	<500 °C
Conductivity			
Heat	Poor to fair	Good to excellent	Poor
Electricity	Nil to some	Excellent	Nil
Coefficient of expansion (thermal)	Low	High	Very high
Max. use temperature	2760 °C	816 °C	260 °C
Water absorption	Some	Nil	some
Electrical insulation	Good to excellent	Nil	Good to excellent

Continued Table 3.3

Property	Ceramic	Metals	Plastlcs
Chemical			
Crystallinity	Usually	Usually	Most are not
Chemical resistance	Excellent	Poor to good	Good to excellent
Composition (raw materials)	Abundant in nature	Some are costly	From oil
Microstructure	From single phase to multicomponent	From single phase to multicomponent	Usually single phase
Fabricability			
Sheet material	Nil	Excellent	Good
Castings	Nil	Excellent	Fair
Extrusions	Nil	Excellent	Good
Machinability	Nil	Good to Excellent	Good
Molding	Fair	Nonferrous only	Excellent

Mechanical Properties

It is extremely difficult to measure the tensile strength of ceramics because the dog-bone-shaped fabrication processes (diamond grinding and the like) and because it is difficult to grip tensile coupons in the machine because of their high hardness. It commonly costs from $600 to $800 to make a single ceramic test coupon. For these reasons, performing tensile tests on ceramics is not normally done, and the available data are often based on fewer replicate samples than one would like. There are ceramics fibers with tensile strengths that can be in excess of 3.4 GPa, but in general the tensile strengths of engineering ceramics are less than 0.69 GPa, much less than some metals but better than most plastics. The property that is often substituted for tensile strength is the four-point ***flexural strength*** test. This test uses a cheaper rectangular bar with dimensions of about 3mm×5mm×50mm, and it is bent as a beam until it breaks. The tensile stress at the outer fibers of the beam at fracture is calculated, and this is used as a measure of tensile strength. It is usually reported as flexural strength. If the ***deflection*** of the beam is measured in this test, these data can be used to produce a stiffness parameter, ***flexural modulus***. This property is often substituted for the tensile modulus of elasticity.

Ceramics do not have a yield strength. They do not plastically deform before reaching their tensile strength. There is no permanent elongation or reduction in area. This is the property that prevents ceramics from competing with metals and plastics for the market that requires forming of materials by room temperature plastic deformation: deep drawing, bending, forming, and the like. These shaping processes simply cannot be used.

The ***compressive strength*** of engineering ceramics can be excellent—better than the compressive strengths of metals, plastics, and composites. Ceramics have the highest hardness of the engineering materials. In fact, diamond (which is considered by most as a ceramic) has the highest hardness of any material that we know. The high hardness of ceramics makes some of them suitable for use as tools for working other materials and for use components.

We have mentioned how hard it is to measure the elastic modulus of ceramics, but it is done and ceramics come out as our stiffest engineering materials. Cemented carbides are three times as stiff as steel; silicon carbide, boron carbide, and titanium diboride are more than twice as stiff as steel. This area is indeed an asset of ceramics as engineering materials. When a steel member is deflecting too much in service and the size cannot be changed, the only solution to the problem is to use a material with greater stiffness than the steel. This leaves few choices but ceramics and composites with high modulus reinforcement.

The long-term service characteristics of ceramics—***fatigue resistance*** and creep strength—are good if the parts are made and loaded in modes favorable to ceramics. The tensile fatigue strength of ceramics is poor. This is because the tensile strength of ceramics is in general poor, and a good design would avoid loading a ceramic in tension. The compressive strength of ceramics is very good. The same sort of situation exists in creep and stress rupture; if the ceramic is loaded in compression, the creep and stress rupture characteristics will be excellent. In fact, ceramics can operate at temperatures of 2000 °F (1093 °C) in furnaces.

In summary, the mechanical properties of ceramics are good in compression, and they can have some special mechanical properties: very high hardness and greater stiffness than other engineering materials. Their greatest weakness is brittleness, and this means that they should be used in design very much the same way that glass is used. Most people can visualize what this means: no stress concentrations, mount in ***resilient material***, do not bend, and so on. If this is done, the strengths of ceramics can be used without getting failures from their weaknesses.

Physical Properties

The density of some of the engineering ceramics is a physical property of concern as well as a physical property strength. Some engineering ceramics are lightweight compared to metals (silicon carbide and boron carbide, for example), and for this reason these materials have specific stiffness that is larger than that of even the highest-strength steels. On the other hand, reduced weight can be an indicator of porosity, which lowers the strength of ceramics. Pressureless sintering of ceramics can produce a material with only 95% of theoretical density. There are two points that we are trying to make: ceramics can be lightweight compared to metals, but low density may also be an indicator of too much porosity in a particular grade. The designer should select a ceramic with high theoretical density (99%) if structural strength is a service consideration.

The melting points of ceramic materials are among the highest of all engineering materials. This is one of their strengths, and it is a reason why ceramics can be used for ***crucible***s and refractory bricks.

The thermal conductivity and electrical conductivity of ceramics cover a large spectrum. The superceramic diamond has the highest thermal conductivity, while most ceramics have low thermal conductivity; they are heat insulators. Low thermal conductivity usually produces poor thermal shock resistance. This is a limiting property for some applications. Most ceramics are electrical insulators, but some are semiconducting. In general, ceramics compete with plastics as electrical insulators, but ceramics take the lead when operating temperatures are in excess of 400 °F (204 °C). Plastic cannot take high temperatures or electric are damage. Ceramics outperform other materials in high-temperature insulation, but some materials such as silicon carbide are actually fair conductors of electricity at elevated temperatures. The point here is that the electrical and thermal properties of each ceramic should be checked at the temperature of interest to avoid surprises.

With regard to thermal expansion, the properties of ceramics cover a range, but they tend to be lower in expansion than metals and plastics. There are glasses that have zero as their coefficient of thermal expansion.

The maximum use temperatures of ceramics are above those of most other engineering materials. This is indeed one of their strengths.

Water absorption is a property that mostly affects plastics, but it can be a factor with ceramics that contain porosity. Water absorption is not a problem with most ceramics that are near theoretical density. Absorbed moisture can lower electrical insulating properties of porous ceramics.

New Words and Expressions

flexural strength ['flekʃərəl streŋθ] 挠（弯）曲强度
deflection [dɪ'flekʃn] *n.* 偏斜，偏转，偏差
flexural modulus ['flekʃərəl 'mɔdjuləs] 挠（弯）曲模量
compressive strength 抗压强度

fatigue resistance [fə'ti:g ri'zistəns] 抗疲劳性
resilient material 弹性材料
crucible ['kru:sɪbl] *n.* 坩埚
Water absorption ['wɔ:tə əb'sɔ:pʃən] 吸水性

Reading Material

Properties of Ceramics (Ⅱ)

Chemical Properties

The engineering ceramics are crystalline; some ceramic-type materials such as glasses are amorphous, not crystalline at all. The ***intermetallic compounds*** and metalloids are, in general, crystalline. Most ceramics do not respond to the types of heat treatments that are used on metals to change crystalline states; however, alloying, the adding of impurity elements, is often used by ceramists to change crystal structure.

The chemical resistance of ceramics is one strength of this class of materials. The driving force for corrosion is that materials want to return to the state in which they were found in nature. Since ceramics are often oxides, nitrides, or sulfides, compounds found in nature, there is little driving force for corrosion. The engineering ceramics, aluminum oxide, silicon carbide, ***zirconia***, silicon nitride and others, are very resistant to chemical attack in wide variety of solutes. The corrosion resistance of ceramics is not always good, however. Corrosion data must be consulted for each ceramics have corrosion characteristics similar to glasses. Glasses are resistant to most acids, bases, and solvents, but a few things like ***hydrofluoric acid*** will rapidly attack them. Corrosion data must be consulted.

It is not common practice to state the chemical composition of ceramics as is done for metals. Where we purchase a steel alloy, it is common practice to ask for a certificate of analysis that shows the percentages of elements present (C, Cr, Ni, Mo, and so on). This is not normally done with ceramics. A drawing on a ceramic part would probably state something like the following: aluminum oxide, 99% theoretical density min. Kors grade K33 or equivalent. This specification would not show the percentages of aluminum and oxygen that compose the aluminum oxide ceramic. Some engineering ceramics are made special by processing during manufacture; a manufacturer may make a blend of aluminum oxide and magnesium oxide that has properties that are quite different from pure aluminum oxide. In these instances, there is little recourse but to specify by trade name. the same thing is true with manufacturer processing that involves special treatments of a ceramic. For example, silicon carbide produced by chemical vapor deposition can have different use properties than the same material made by conventional pressing and sintering. Hipping, ***hot isostatic pressing***, of ceramics improves ***densification***, and this type of processing can alter use properties. The microstructure and porosity can vary with the method of manufacture. Thus it is not common to specify chemical composition in specifying ceramics, but it is recommended that density, processing, and additives be part of a specification. Very often ceramics are specified by trade name and manufacturer. In 2000 this was about the only way that you could be assured of getting the same material on each other.

Fabricability

Ceramics would be used much more in machine design and other industrial applications if it were not for their fabrication limitations. If a part is made by sheet metal fabrication processes, these processes will not work at all with ceramics. They cannot be plastically deformed to shapes at room temperatures. Ceramics cannot be melted and cast to shapes like metals, mostly because of their high melting temperatures. Silicon carbide melts at 4700 °F (2600 °C); aluminum oxide melts at 3659 °F (2015 °C). Ceramics types of materials are normally used for the molds for casting metals. A new form of material would have to be developed for molds for ceramic materials. There are additional problems in casting ceramics: the furnaces needed to melt a material with a melting point in excess of 3000 °F (1648 °C). Normal furnaces cannot achieve this type of temperature. Melting would have to be done with are processes or something like an electron beam. The same situation exists with extrusions. What can be used for tooling to extrude a material at temperatures in excess of 3000 °F (1648 °C)?

The machining of ceramics is so costly that it is not an option for many machine parts. It was pointed out in our discussion of mechanical properties that tensile test specimens may cost $600 each; the same situation exists for one-of-a-kind parts for machines. It is not uncommon to pay $400 each for a 25-mm-OD, 4-mm-wall, 25-mm-long plain bearing of aluminum oxide. If large quantities of parts are needed and if they are relatively small, it is possible to make these by ***injection molding*** or extrusion processes, and the cost can become very low. Unfortunately, most service applications do not require the 100 000-plus annual part requirement that would justify the necessary tooling.

Probably the best example of the successful use of ceramics for a low-cost part is the common spark plug used in automobiles. The insulator is aluminum oxide (because ceramics are electrical insulators that can take the temperatures). On sale, spark plugs can be purchased for less than $1 each. The aluminum oxide insulator on the plug is a fairly complicated shape; how can this ceramic part be produced so cheaply? The answer is that the ceramic insulation is injection molded around the metal components in a molding machine that works similarly to those used for plastics. Aluminum oxide in fine powder form is injected into rubber molds at extremely high pressures. No machining is required; parts are molded and sintered to size. In this way, ceramics can be made at low cost. Many electrical parts are made this way, but it is the ***norm*** to have production requirements in excess of 1 million per year to justify the expense of the tooling required for injection molding. Ceramics often shrink 30% on sintering. To get a part to come out to the right dimensions after sintering, a shrink allowance must be put in the mold; this is the engineering aspect of molding ceramics to size.

In summary, in 2000 most ceramics require expensive processing to bring them to a usable shape. The low-cost processes are injection molding, extrusion, and pressureless sintering. High performance engine parts and similar structural parts require expensive tooling and processing that are difficult to justify on small quantities of parts. The sol-gel process may someday allow these limitations to be overcome. Parts can be formed in gels that behave like plastics, and sintering converts them to useful ceramics parts. This technology does not exist on a commercial basis in 2000, but it may come.

New Words and Expressions

intermetallic compound [ˌintə(:)mi'tælik 'kɔmpaund] 金属间化合物，金属互化物
zirconia [zə'kəunɪə] *n.* 氧化锆
hydrofluoric acid ['haidrəflu(:)'ɔrik 'æsid] *n.* 氢氟酸
densification [densɪfɪ'keɪʃən] *n.* 致密，捣实，增浓作用，稠化（作用）
hot isostatic pressing 高温等静压，热等静压
injection molding [in'dʒekʃən 'məuldiŋ] 喷射造型法，喷射模塑法
norm [nɔ:m] *n.* 标准，规范

Lesson 10 Electronic Ceramics—Electrical Insulators and Conductors

There are literally hundreds of applications of advanced ceramics that depend primarily on the reaction of the material to applied electric or magnetic fields. Some of these are enumerated here, along with a brief description of the special characteristics that make these materials useful for particular applications. In many cases, while the electronic properties are paramount, for many of these applications there are also stringent mechanical and thermal property requirements that must be met.

Many ceramic materials are electrical insulators and, consequently ceramics have been used for years for ***dc*** and low-frequency ac electrical insulator shapes ranging from large, high-voltage suspension insulators for power transmission lines to simple, low-voltage shapes for lamp and switch bases. These shapes have traditionally been made from clay-based porcelains and are not usually included in the advanced ceramics category. On the other hand, the utilization of advanced ceramic electrical insulator materials suitable for more exotic applications is growing very rapidly. The materials most often used are alumina ceramics, beryllia ceramics, aluminum nitride, and a variety of special glasses, including those that can be converted into crystalline form after shaping (glass-ceramics). The most important electrical properties of such insulation materials are very low electrical conductivity, low ***dielectric constant*** (a low tendency to ***polarize*** or store charge), a high dielectric strength (resistance to breakdown under large voltage drops), and, for high frequency applications, low ***dielectric losses*** (low propensity to convert energy in the alternating field into heat).

A wide variety of shapes are made from advanced ceramic insulator materials, many of which are so intricate that they must be made by injection molding or isostatic pressing followed by machining and finishing. An especially important ceramic insulation application is as smooth substrates for thick film and thin film ***deposition*** of circuitry. Substrates are usually made as thin (a few mm), flat rectangular sheets utilizing tape-casting technology. Most frequently, discrete electronic devices such as silicon chips or discrete capacitors will be attached to the film circuitry on the substrate to form what are known as hybrid circuits. It is not at all unusual for multilayer ceramic substrates to be employed. Multilayer substrates are made by thick-film printing of circuitry onto unfired ceramic tapes using metal inks, then stacking and laminating the green tapes together to form a sandwich structure, and then "cofiring" the ceramic and metal inks to form a single mutilayer substrate. The circuits on different layers of the multilayer structure are connected at appropriate points by metal-filled holes called vias in the intervening ceramic layers. Substrates not only support the circuitry, but they also provide for dissipation of heat generated in the circuitry, either by absorbing it themselves or by conveying it to an attached heat sink. When substrates and their associated circuitry are fitted with external leads and are encapsulated to protect the circuitry from the environment, the entire assembly is usually called an electronic package.

Ceramic insulator materials are also commonly used as capacitor dielectrics that is , as the material placed between the plates of a ***capacitor*** to serve as the ***charge storage medium***. While any insulating material can be used for such an application, it is usually desirable to use materials that will allow the maximum amount of charge storage (capacitance) in the smallest possible device. This consideration means that materials with very high dielectric constants should be used. In addition to high dielectric constant, a capacitor dielectric should have high dielectric strength and low dielectric losses and should exhibit minimal variations in these properties with temperature or voltage changes. The most important group of advanced ceramic capacitor dielectric materials consists of combinations of barium titanate ($BaTiO_3$) with a variety of other oxides used to modify its fundamental properties. There are hundreds of titanate-based materials in use. Ceramic capacitor dielectric are often made in the form of small, thin discs or thin-walled hollow tubes, with the plates being deposited on each side by thick film techniques.

A very important and rapidly growing form of high-rating ceramic capacitor, called a ceramic chip capacitor, is made by a process similar to that used for multilayer ceramic substrates. Very thin sheets of titanate dielectric are produced by tape casting, and a pattern of metal electrodes is thick-film printed onto one side. Many layers of tape are then stacked on top of one another and laminated together. Individual "chips" are diced out of this ***laminate*** and are fired to mature the ceramic-metal sandwich. These tiny chip capacitors can be soldered directly onto printed circuitry.

A number of ceramic materials are electrical insulators with respect to the movement of electrons, nevertheless they exhibit measurable electrical conductivities because of the ability of certain ions to move through the material when an electric field is applied. Such materials are called ionic conductors. If the conductivity is relatively high, they are called fast ion conductor or solid electrolytes. The most important fast ion conductors are AgI (Ag is the conducting ion), CaF_2 (F is the conducting ion), the so-called beta-aluminas (having roughly the formula $MAl_{11}O_{17}$, where M is silver or an alkali such as sodium, the M ion being the one responsible for conduction), zirconia (ZrO_2) doped with lime or yttrium oxide (with O being the conducting ion), and a number of special glasses (usually with ***alkali*** ions imparting conduction). Generally, the conductivity of ionic conductors increases rapidly with an increase in temperature, so they are almost always utilized at temperatures above room temperature, and sometimes at quite high temperatures. Their behavior as purely ionic conductors allows their use as solid electrolytes in high-temperature ***batteries*** and fuel cells, and the fact that only one particular type of ion moves in an electric field makes them useful as ion-specific sensor materials (an example is the use of stabilized zirconia as an oxygen sensor in automobile exhaust systems to sense the efficiency of the combustion process and activate changes in fuel-to-air ratios).

Although silicon, germanium, and ***gallium arsenide*** are the most utilized semiconductor materials, a number of other ceramic materials also are employed for semiconductor applications. Among the most used for these applications are various doped or slightly reduced oxides (especially ZnO) and doped silicon carbide. Such materials are commonly used as varistors (resistance changes with applied voltage) and thermistors (resistance changes with temperature). Varistors are commonly used to protect devices from damage by line surges (such as may be caused by lightning) or by shunting current around the device when the ***varistor*** becomes highly conductive due to a voltage spike. ***Thermistors*** can be used as temperature measurement devices, and, if they are doped so as to have a positive temperature coefficient of resistance (their resistance increases with increasing temperature), they can be used as self-limiting heater elements in a variety of applications, including to rapidly heat automatic chock elements in automobile engines so that the chock quickly closes after start-up. When fabricated into single crystal form, ceramic semiconducting materials can be used to form pn junction diodes, and these can be used as power transistors, as ***light-emitting diodes*** (LEDs), and even as semiconductor laser diodes.

New Words and Expressions

dielectric constant [ˌdaii'lektrik 'kɔnstənt] 介电常数
polarize ['pəʊləraɪz] *v.*（使）极化，（使）偏振化
dielectric losses 电介质损失
deposition [ˌdepə'zɪʃn] *n.* 沉积（物）
laminate ['læmɪnət] *n.* 层压材料，叠层，层压
capacitor [kə'pæsɪtə(r)] *n.* 电容器
alkali ['ælkəlaɪ] *n.* 碱；*adj.* 碱性的
batteries ['bætərɪz] *n.* 电池
gallium arsenide ['gæliəm 'ɑːsənaid] 砷化镓
varistor [væerɪstə] *n.* 压敏电阻，可变电阻
thermistor [θɜː'mɪstə] *n.* 热敏电阻
light-emitting diodes 发光二极管

Reading Material

Applications of Ceramics

Traditional ceramics are quite common, from sanitary ware to fine Chinese porcelains to glass products. Currently ceramics are being considered for uses that a few decades ago were inconceivable; applications ranging from ceramic engines to optical communications, ***electrooptic*** applications to laser materials, and substrates in electronic circuits to electrodes in ***photoelectrochemical devices***. Some of the recent applications for which ceramics are used and/or are prime candidates are listed in Table 3.4.

Historically, ceramics were mostly exploited for their electrical insulative properties, for which ***electrical porcelains*** and aluminas are prime examples. Today, so-called electrical and electronic ceramics play a pivotal role in any modern technological society. For example, their insulative properties together with their low-loss factors and excellent thermal and environmental stability make them the materials of choice for substrate materials in electronic packages. The development of the perovskite family with exceedingly large dielectric constants holds a significant market share of capacitors produced. Similarly, the development of magnetic ceramics based on the spinel ferrites is today a mature technology. Other electronic/electrical properties of ceramics that are being commercially exploited include piezoelectric ceramics for sensors and actuators, nonlinear *I-V* characteristics for circuit protection, and ***ionically conducting ceramics*** for use as solid electrolytes in high-temperature fuel cells and as chemical sensors.

These applications do not even include superconducting ceramics, currently being developed for myriad applications.

Mechanical applications of ceramics at room temperature usually exploit hardness, wear, and corrosion resistance. The applications include cutting tools, ***nozzles***, ***valves***, and ball bearings in aggressive environments. However, it is the refractoriness of ceramics and their ability to sustain high loads at high temperatures, together with their low densities, that has created the most interest. Applications in this area include all ceramic engines for transportation and turbines for energy production. In principle, the advantages of an all-ceramic engine are several and include lower weight. higher operating temperatures which translates to higher efficiencies, and less pollution. It is also envisioned that such engines would not require cooling and maybe not even any ***lubrication***, which once more would simplify the design of the engine, reducing the number of moving parts and lowering the overall weight of the ***vehicle***.

Table 3.4 Properties and applications of advanced ceramics

Property	Applications (examples)
Thermal	
Insulation	High-temperature furnace linings for insulation (oxide fibers such as SiO_2, Al_2O_3, and ZrO_2)
Refractoriness	High-temperature furnace linings for insulation and containment of molten metals and slags
Thermal conductivity	Heat sinks for electronic packages (AIN)
Electrical and dielectric	
Conductivity	Heating elements for furnaces (SiC, ZrO_2, $MoSi_2$)
Ferroelectricity	Capacitors (Ba-titanate-based materials)
Low-voltage insulators	Ceramic insulation (porcelain, steatite, forsterite)
Insulators in electronic applications	Substrates for electronic packaging and electrical insulators in general (Al_2O_3, AlN)
Insulators in hostile environments	Spark plugs (Al_2O_3)
Ion-conducting	Sensor and fuel cells (ZrO_2, Al_2O_3, etc.)
Semiconducting	Thermistors and heating elements (oxides of Fe, Co, Mn)
Nonlinear *I-V* characteristics	Current surge protectors (Bi-doped ZnO, SiC)
Gas-sensitive conduct	Gas sensors (SnO_2, ZnO)
Magnetic and superconductive	
Hard magnets	Ferrite magnets [(Ba, Sr) O · $6Fe_2O_3$]
Soft magnets	Transformer cores [(Zn, M) Fe_2O_3, with M = Mn, Co, Mg]; magnetic tapes (rare-earth garnets)
Superconductivity	Wires and SQUID magnetometers ($YBa_2Cu_3O_7$)
Optical	
Transparency	Windows (soda-lime glasses), cables for optical communication (ultra-pure silica)
Translucency and chemical inertness	Heat- and corrosion-resistant materials, usually for Na lamps (Al_2O_3MgO)
Nonlinearity	Switching devices for optical computing ($LiNbO_3$)
IR transparency	Infrared laser windows (CaF_2, SrF_2, NaCl)
Nuclear applications	
Fission	Nuclear fuel (UO_2, UC), fuel cladding (C, SiC), neutron moderators (C, BeO)
Fusion	Tritium breeder materials (zirconates and silicates of Li, Li_2O); fusion reactor lining (C, SiC, Si_3N_4)
Chemical	
Catalysis	Filters (zeolites); purification of exhaust gases
Anticorrosion	Heat exchangers (SiC), chemical equipment in corrosive environments
Biocompatibility	Artificial joint prostheses (Al_2O_3)
Mechanical	
Hardness	Cutting tools (SiC whisker-reinforced Al_2O_3, Si_3N_4)
High-temperature strength retention	Stators and turbine blades, ceramic engines (Si_3N_4)
Wear resistance	Bearings (Si_3N_4)

New Words and Expressions

electrooptic *adj.* 电光的

photoelectrochemical *adj.* 光电化学的

electrical porcelains 电瓷

ionically conducting ceramics 离子导电陶瓷

nozzle ['nɒzl] *n.* 管口，喷嘴

valve [vælv] *n.* 阀，真空管

lubrication [ˌluːbrɪ'keɪʃn] *n.* 润滑

vehicle ['viːəkl] *n.* 交通工具，车辆

UNIT 4 POLYMERS

Lesson 11 Introduction to Polymers

The term polymer comes from "poly," meaning many, and "mer," meaning units. Hence, polymers are composed of many units—in this case, structural units called ***monomers***. A monomer is any unit that can be converted into a polymer. Similarly, a ***dimer*** is a combination of two monomers, a ***trimer*** is a combination of three monomers, and so on. polymer chains are formed from the reaction of many monomers to form long chain ***hydrocarbons***, sometimes called macromolecules, but more commonly referred to as polymers.

Polymer Classification

It is useful to classify polymers in order to make generalizations regarding physical properties, formability, and reactivity. The appropriate classification scheme can change, however, because there are several different ways in which to classify polymers. The first scheme groups polymers according to their chain chemistry. Carbon chain polymers have a backbone composed entirely of carbon atoms. In contrast, heterochain polymers have other elements in the backbone, such as oxygen in a ***polyether***, —C—O—C—. We can also classify polymers according to their macroscopic structure—that is, independent of the chemistry of the chain or ***functional groups***. There are three categories of polymers according to this scheme: linear, branched, and networked (***crosslinked***) polymers. Refer to Figure 4.1 for a schematic representation of this classification scheme.

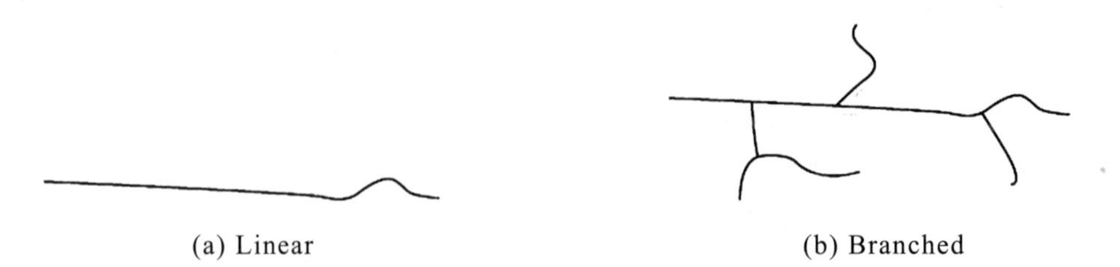

(a) Linear (b) Branched

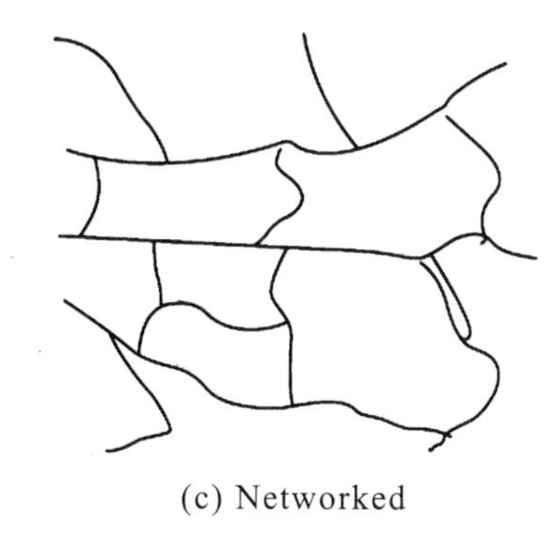

(c) Networked

Figure 4.1 Classification of polymers according to macroscopic structure

Finally, polymers can be classified according to their formability. Polymers that can be repeatedly shaped and reshaped are called thermoplastics, whereas those polymers that cannot be reshaped at any temperature once they are set are termed thermosets. Often times network polymers are thermosets, and linear and branched polymers are thermoplastics. Hence, the thermoplastic/thermoset distinction is worthy of some elaboration.

Thermoplastic Polymers

Most thermoplastic polymers are used in high volume, widely recognized applications, so they are often referred to as commodity plastics. (we simply note that a plastic is a polymer that contains other additives and is usually identified by a variety of commercial trade names. There are numerous databases, both in books and on the internet, that can be used to identify the primary polymer components of most plastics. With a few notable exceptions, we will refer to most polymers by their generic chemical name.) The most common commodity thermoplastics are polyethylene (PE), ***polypropylene*** (PP), ***polyvinyl chloride*** (PVC) and ***polystyrene*** (PS). These thermoplastics all have in common the general repeat unit —(CHX—CH_2)—, where —X is —H for PE, —CH_3 for PP, —Cl for PVC, and a benzene ring for PS. When we discuss ***polymerization reactions*** in next unit, we will see that all of these thermoplastics can be produced by the same type of reaction. In their simplest forms, the thermoplastics are linear, carbon chain polymers. There are methods for creating branches, especially in polyethylene, while still maintaining thermoplasticity. Increased branching tends to decrease the density, melting point, and certain mechanical properties of the polymer, but increases transparency and impact toughness. Thus, branched PE is important for many packaging applications. Other special types of PE include low-density PE (LDPE), high-density PE (HDPE), and linear low-density PE (LLDPE). HDPE is used when greater stiffness is required, such as in milk, water and detergent bottles. LDPE is used for many standard piping applications. LLDPE has a low density like LDPE, but a linear structure much like HDPE. It is less expensive to produce than LDPE, and it generally has better mechanical properties than LDPE. PP, PVC, and PS have in common an asymmetric carbon atom in their backbone, a fact that leads to interesting structural properties. They all have specific advantages leading to a variety of applications. For the moment, it is important to recognize that they are linear thermoplastics, whose properties can be tailored through blending, branching, and additives. It is important to note at this point that the ability of thermoplastics to soften when heated and harden upon cooling is what leads to the principle of recycling.

Thermoset Polymers

There are many important examples of thermoset polymers, a subset of which are sometimes referred to as resins. The thermoplastic PE can be treated by electron radiation or chemical means to form chemical bonds between adjacent chains called crosslinks. In this process, some of the carbon-hydrogen or carbon-carbon bonds in the linear chain are broken, creating ***free radicals***, which react with free radicals on other chains to form bridges between the chains. When the crosslinking is brought about by chemical means, the term curing is often employed. The result is that unlike linear PE, crosslinked PE has no distinct melting point due to limited chain mobility and will eventually degrade upon heating. This is a common characteristic of thermoset polymers. Due to their three-dimensional structure, they cannot be reshaped upon heating. Instead, they tend to remain rigid, or soften only slightly, until the backbone begins to break, leading to polymer ***degradation***. While this may not seem like a desirable quality, it means that thermoset polymers are generally stiffer than thermoplastics. In fact, the crosslinks tend to increase the long-term thermal and dimensional stability of the polymer, such that thermosets find wide use in very large and complex parts where thermal stability is important. The chemical structures of thermosets are generally much more diverse than the commodity thermoplastics. The most common types of thermosets are the ***phenol-formaldehydes*** (PF), ***urea-formaldehydes*** (UF), ***melamine-formaldehydes*** (MF), epoxies (EP), ***polyurethanes*** (PU), and ***polyimides*** (PI). A related, yet distinctly different, class of crosslinked polymers are the elastomers. Though they are structurally different than the thermosets, we will include them here since they tend to decompose when heated, rather than flow. The presence of crosslinks in some polymers allows them to be stretched, or elongated, by large amounts. Polymers that have more than 200% elastic elongation (three times the original length) and can be returned to their original form are termed elastomers. They are like thermoplastics in that they readily elongate, but the presence of crosslinks limits the elongation prior to breakage and allows the polymer to return to its original shape. Natural rubber is an important elastomer. Crosslinks are added to an emulsion of rubber, called latex, through the addition of heat and sulfur. The sulfur creates chemical bonds between the rubber chains in a process known as ***vulcanization***. Other common elastomers are ***polyisoprene***, ***butadiene rubber*** (BR), ***styrene butadiene rubber*** (SBR), ***silicones***, and ***fluoroelastomers***.

New Words and Expressions

monomer ['mɒnəmə] *n.* 单体
dimer ['daɪmə] *n.* 二聚物
trimer ['traɪmə] *n.* 三聚物
hydrocarbon [haɪdrə'kɑːbən] *n.* 碳氢化合物，烃
polyether [pɒlɪ'iːθə] *n.* 聚醚
functional groups ['fʌŋkʃənəl gruːp] 官能团
crosslinked [k'rɒslɪŋkt] *adj.* 交联的
polypropylene (PP) [ˌpɒli'prəʊpəliːn] *n.* 聚丙烯
polyvinyl chloride (PVC) [ˌpɔli'vainil 'klɔːraid] *n.* 聚氯乙烯
polystyrene (PS) [ˌpɒli'staɪriːn] *n.* 聚苯乙烯
polymerization reaction 聚合反应
free radical [friː 'rædikəl] 自由基，游离基
degradation [ˌdegrə'deɪʃn] *n.* 降解
phenol-formaldehyde (PF) ['fiːnəl fɔː'mældəˌhaɪd] *n.* 酚醛树脂
urea-formaldehyde (UF) [jʊə'rɪəfɔːm'ældɪhaɪd] *n.* 脲醛树脂
melamine-formaldehyde (MF) ['meləmiːn fɔː'mældɪhaɪd] *n.* 三聚氰胺甲醛树脂，蜜胺树脂
polyurethane (PU) ['pɑːli'jɜrəˌθeɪn] *n.* 聚氨基甲酸酯，简称聚氨酯
polyimide (PI) [pɒlɪ'ɪmaɪd] *n.* 聚酰亚胺
vulcanization [ˌvʊlkənaɪ'zeɪʃən] *n.* （橡胶的）硫化（过程），硫化
polyisoprene [pɒlɪ'aɪsəʊpriːn] *n.* 聚异戊二烯，天然橡胶
butadiene rubber [ˌbjuːtə'daiiːn 'rʌbə] *n.* 丁二烯橡胶，聚丁橡胶
styrene butadiene rubber (SBR) ['stairiːn ˌbjuːtə'daiiːn 'rʌbə] *n.* 丁苯橡胶
silicone ['sɪlɪkəʊn] *n.* 硅树脂，（聚）硅酮
fluoroelastomer [fluːərəɪ'lɑːstəmə] *n.* 含氟弹性体，氟橡胶

Reading Material

Reviews of Materials Physics: Polymers

Polymers, also known as macromolecules, are long chain molecules in which a molecular unit repeats itself along the length of the chain. The word polymer was coined approximately 165 years ago (from the Greek polys, meaning many, and meros, parts). However, the verification of the structure of polymers, by diffraction and other methods, had to wait, approximately, another 100 years. We now know that the DNA molecule, ***proteins***, ***cellulose***, silk, wool, and rubber are some of the naturally occurring polymers. Synthetic polymers, derived mainly from petroleum and natural gas, became a commodity starting approximately 50 years ago. Polymers became widely known to the public when nylon was introduced as a substitute for silk and, later, when ***Teflon***-coated pans became commercially available. Polymers are now widely used in numerous household applications. Their industrial use is even more widespread.

Most of the applications associated with polymers have been as structural materials. Since the 1970s it was realized that with suitable doping of the polymers, a wide variety of physical properties could be achieved, resulting in products ranging from ***photosensitive materials*** to superconductors. The field of materials physics of polymers has grown rapidly from this period onwards.

Polymers are a remarkably flexible class of materials, whose chemical and physical properties can be modified by molecular design. By substitution of atoms, by adding side groups, or by combining (blending) different polymers, chemists have created a myriad of materials with remarkable, wide ranging, and useful properties. This research is largely driven by the potential applications of these materials in many diverse areas, ranging from cosmetics to electronics. Compared to most other materials, polymers offer vast degrees of freedom through blending and are generally inexpensive to fabricate in large volumes. They are light weight and can have very good strength-to-weight ratios.

Polymers have traditionally been divided into five classes:

(1) Plastics are materials that are molded and shaped by heat and pressure to produce low-density, transparent, and often tough products, for uses ranging from beverage bottles to shatterproof windows.

(2) Elastomers are chemically cross-linked or entangled polymers in which the chains form irregular coils that straighten out during strain (above their glass transition temperatures), thus providing large elongations, as in natural and synthetic rubbers.

(3) Fibers, which are spun and woven, are used primarily in fabrics. About fifty

million tons of fibers are produced annually for uses ranging from clothing to drapes. Apart from naturally occurring fibers such as silk and wool, there are regenerated fibers made from cellulose polymers that make up wood (rayon) and synthetic fibers, comprising molecules not found in nature (nylon).

(4) Organic adhesives have been known since antiquity. However, with demanding environments and performance requirements, synthetic adhesives and glues have largely replaced natural ones. The microscopic mechanisms of adhesion and the toughness of joints are still debated. There is an increasing trend to use UV radiation to promote polymerization in adhesives and, more generally, as a method of polymerization and cross linking in polymers.

(5) Finally, polymers, frequently with additives, are used as protective films, such as those found in paints or varnishes.

Physicists have played a significant role in explaining the physical properties of polymeric materials. However, the interest of physicists in polymers accelerated when it was discovered that ***polyacetylene*** could be made conductive by doping. This development was noteworthy for it opened the possibility of deliberately controlling conductivity in materials that are generally regarded as good insulators. The structure of all ***conjugated polymers***, as these materials are known, is characterized by a relatively easily delocalized p bond, which, with suitable doping, results in effective charge motion by solitons, polarons, or bipolarons. Since the discovery that polymers could be electrical conductors, active research areas have developed on the physics of polymer superconductors, ferro- and ferri-magnets, piezoelectrics, ferroelectrics, and ***pyroelectrics***. Within the field of doped polymers, devices have been built to demonstrate light-emitting diodes, ***photovoltaic cells***, and transistors.

Conjugated polymers have also been investigated extensively for their large nonlinear, third-order polarizability, which is of interest to the field of nonlinear optics. Large nonlinearities are associated with the strong polarizability of the individual molecules that make up the building blocks of the polymer. Furthermore, the flexibility of polymer chemistry has allowed the optical response of polymers to be tailored by controlling their molecular structure, through the selective addition of photoactive molecules. Hence these materials have been widely investigated by physicists and engineers for optical applications, such as in holographic displays (dichromated gelatin), diffraction gratings, ***optocouplers***, and wave guides.

Polymers have long interested physicists for their conformational and topological properties. This interest has shifted from the conformational behavior of individual molecules to that of a macromolecular assembly, phase behavior, and a search for universal classes. ***Block copolymers***, consisting of two or more polymers, can give rise to nanoscale phases, which may, for example, be present as spheres, rods, or parallel lamellae. The distribution of these phases and their ***topologies*** are of current theoretical and practical interest. Block copolymer ***morphologies*** are also being used as nanoscale templates for production of ceramics of unique properties having the same morphology.

Block copolymers are also of interest as biomaterials. Proteins are an example of block copolymers, in which the two phases form helical coils and sheets. Attempts to mimic the ***hierarchical structure*** present in natural polymers have only been partly successful. The principal difficulty has been to control the length of the polymer chains to the precision that Nature demands. Significant progress has been made in controlling polymer morphologies with the use of new catalysts. For example, ***metallocenes*** have been used as catalysts to control branched polymers and organonickel ***initiators*** to suppress ***chain transfer*** and ***termination***, so that ***polypeptides*** with well-defined sequences and with potential for applications in tissue engineering could be made. The growth of well-controlled polymer chains is an example of "living" polymerization.

The static and dynamic arrangement of atoms on the surfaces and interfaces of polymers is another area of active investigation. For example, thin films of polymers, in which the chain lengths are long compared to the thickness of a film, show unusual physical properties: the glass transition temperature for a thin-film polymer decreases significantly, but between solid surfaces polymer liquids solidify.

Even though we have some way to go in making tailored proteinlike structures, polymer research has played a significant role in the class of materials called biomaterials. Polymers have been used, for example, to produce artificial skin, for dental fillings that are polymerized in situ by a portable UV lamp, and for high density polyethylene used in knee prostheses. Physicists play a significant role in these developments, not only for their interest in the materials, but also because of their familiarity with physical processes that can be used to tailor the properties of polymers. A particularly good example of this interplay is the recent and rapidly growing use of ***excimer laser*** radiation to correct corneal abnormalities; using a technology developed from studies of the ablation of polymeric materials for applications in the electronics industry, physicists realized that the small, yet precisely controlled, ablation of a polymeric surface might be useful in shaping the surface of an eye.

New Words and Expressions

protein ['prəʊtiːn] *n.* 蛋白质

cellulose ['seljuləʊs] *n.* 纤维素

Teflon ['teflɒn] *n.* 特氟纶，聚四氟乙烯（商标名称）

photosensitive materials [ˌfəʊtəʊ'sensətɪv mə'tɪərɪəlz] 感光材料，光敏材料

polyacetylene [pɒlɪə'setəliːn] *n.* 聚乙炔

conjugated polymer ['kɔndʒugeitid 'pɔləmə] 共轭聚合物，共轭高分子

pyroelectric [paɪrəʊɪ'lektrɪk] *adj.* 焦热电的，热释电的

photovoltaic cell [ˌfəutəuvɔl'teiik sel] 硒电池，光电池

transistor [træn'zɪstə(r)] *n.* 晶体管

optocoupler [ɒptəʊ'kuːplər] *n.* 光耦合器，光电隔离器

block copolymer [blɔk kəu'pɔlimə] 嵌段共聚物

topology [tə'pɒlədʒi] *n.* 拓扑结构

morphology [mɔː'fɒlədʒi] *n.* 形态学

hierarchical structure [ˌhaiə'rɑːkikəl 'strʌktʃə] 层次结构，分级结构

metallocene [mi'tæləsiːn] *n.* 金属茂（合物），茂（合）金属

initiator [i'niʃieitə(r)] *n.* 引发剂

chain transfer [tʃein træns'fəː] 链转移

termination [ˌtɜːmɪ'neɪʃn] *n.* 终止

polypeptide [ˌpɒlɪ'peptaɪd] *n.* 多肽

excimer laser [ek'saimə 'leizə] *n.* 受激准分子激光器

Lesson 12 Polymer Synthesis

The study of polymer science begins with understanding the methods in which these materials are synthesized. Polymer synthesis is a complex procedure and can take place in a variety of ways. It is conventional to divide the synthesis of polymers into two main categories. One is ***step-growth polymerization*** which is often also called ***condensation polymerization*** since it is almost exclusively concerned with condensation reactions taking place between multifunctional monomer molecules. The other category is ***chain (addition) polymerization*** where the monomer molecules add on to a growing chain one at a time and no small molecules are eliminated during the reaction.

Condensation Polymerization

In order to produce polymer molecules by step-growth polymerization it is essential that the monomer molecules must be able to react at two or more sites. In the condensation reaction, each of the two reactants can only react at one point. Because of this once the initial reaction has taken place the ester produced is incapable of taking part in any further reactions. Now, if reactants are used which are capable of reacting at two sites polymerization can start to take place. A reaction between a diacid and diol (di-alcohol) yields an ester which still has unreacted end groups.

In the general reaction it is assumed that n molecules of diacid reacted with n molecules of diol. In fact it is usually necessary that equimolar mixtures (i.e. equal numbers of functional groups) are used otherwise the yield of high-molar-mass polymer is low. This conditions is very difficult to achieve in the laboratory and is especially so in an industrial situation. If there is an excess of diacid, for example, intermediate short polymer molecules are produced which have acid end groups:

$$HO\text{-}[OCR_1COOR_2O]_n\text{-}OCR_1COOH$$

These intermediates are unable to react with further growing chains of the same type and so the reaction is effectively blocked. One of the ways of overcoming the problem is to use an ***ω-hydroxy carboxylic*** acid as monomer rather than the diacid and diol. The ω-hydroxy carboxylic acid molecule, HOOC—R—OH (R—general group) has an acid group at one end and a hydroxyl group at the other. These molecules are able to undergo condensation reactions with others of their own type but they also have the great advantage of, if they are pure, guaranteeing equal numbers of functional groups in the reaction. Two monomer molecules will react to form a dimer

$$HCOO—R—OH + HCOO—R—OH \longrightarrow HCOO—R—OOCR—OH + H_2O$$

Which is also an ω-hydroxy carboxylic acid. The dimers can react with either monomer or dimer molecules and so chain growth can proceed as before.

Another factor which can tend to reduce the yield of long polymer molecules, even when equal numbers of functional groups are employed, is the equilibrium which occurs between the reactants and products during condensation reactions. If the concentration of the condensate (such as water) is allowed to build up the reaction may stop and can even be forced in the reverse direction. This problem is usually readily overcome by removing the concentration of the condensate and so driving the reaction in the forward direction. Also the ***kinetics*** of ***polyesterification*** can be rather slow at ambient temperature and so the reactants are often heated to a higher temperature and a catalyst is sometimes added. Both of these two modification can speed up the reaction.

The principles outlined above for polyesterification reactions are equally applicable to step-growth polymerization reactions involving other types of monomer. For example, diamines and dicarboxylic acids combine together to give ***polyamides*** (nylons). The most notable exception is that certain step-growth polymerization reactions in which polyurethanes are formed do not produce any condensation products.

Addition Polymerization

Addition polymerization is the second main type of polymerization reaction. It differs from step-growth polymerization in several important ways. It takes place in three distinct steps initiation, ***propagation*** and termination and the principal mechanism of polymer formation is by addition of monomer molecules to a growing chain. Monomers for addition polymerization normally contain double bonds and of the general formula $CH_2 = CR_1R_2$. the double bond is susceptible to attack by either free radical or ionic initiators to form a species known as an active centre. This active centre propagates a growing chain by the addition of monomer molecules and the active centre is eventually neutralized by a termination reaction. Since the reaction only occurs at the reactive end of the growing chain long molecules are present at an early stage in the reaction along with unreacted monomer molecules which are around until very near the end of the reaction.

The most common type of addition polymerization is ***free radical polymerization***. A free radical is simply a molecule with an unpaired electron. The tendency for this free radical to gain an additional electron in order to makes it highly reactive so that it breaks the bond on another molecule by stealing an electron, leaving that molecule with an unpaired election (which is another free radical). Free radicals are often created by the division of a molecule (known as an initiator) into two fragments along a single bond. The following diagram shows the formation of a radical from its initiator, in this case ***benzoyl peroxide***.

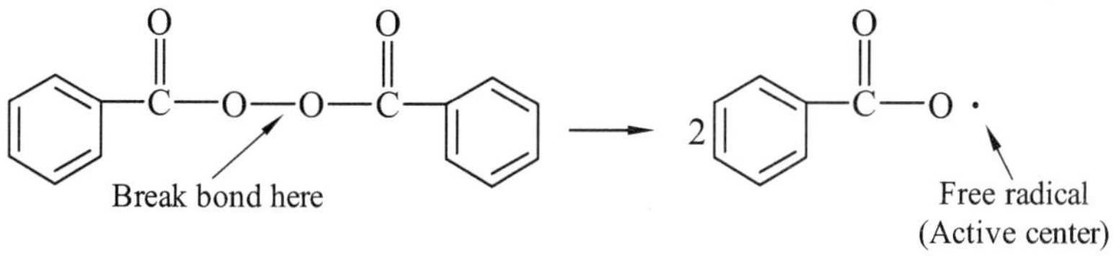

$$C_6H_5-\overset{\overset{O}{\|}}{C}-O-O-\overset{\overset{O}{\|}}{C}-C_6H_5 \longrightarrow 2\,C_6H_5-\overset{\overset{O}{\|}}{C}-O\cdot$$

Break bond here

Free radical
(Active center)

The stability of a radical refers to the molecule's tendency to react with other compounds. An unstable radical will readily combine with many different molecules. However a stable radical will not easily interact with other chemical substances. The stability of free radicals can vary widely depending on the properties of the molecule. The active center is the location of the unpaired electron on the radical because this is where the reaction takes place. In free radical polymerization, the radical attacks one monomer, and the electron migrates to another part of the molecule. This newly formed radical attacks another monomer and the process is repeated. Thus the active center moves down the chain as the polymerization occurs.

There are three significant reactions that take place in addition polymerization: initiation (birth), propagation (growth), and termination (death). These separate steps are explained below.

Initiation Reaction

The first step in producing polymers by free radical polymerization is initiation. This step begins when an initiator decomposes into free radicals in the presence of monomers. The instability of carbon-carbon double bonds in the monomer makes them susceptible to reaction with the unpaired electrons in the radical. In this reaction, the active center of the radical "grabs" one of the electrons from the double bond of the monomer, leaving an unpaired electron to appear as a new active center at the end of the chain. Addition can occur at either end of the monomer.

In a typical synthesis, between 60% and 100% of the free radicals undergo an initiation reaction with a monomer. The remaining radicals may join with each other or with an impurity instead of with a monomer. "self destruction" of free radicals is a major hindrance to the initiation reaction. By controlling the monomer to radical ratio, this problem can be reduced.

Propagation Reaction

After a synthesis reaction has been initiated, the propagation reaction takes over. In the propagation stage, the process of electron transfer and consequent motion of the active center down the chain proceeds. In this diagram, (chain) refers to a chain of connected monomers, and X refers to a ***substituent group*** (a molecular fragment) specific to the monomer. For example, if X were a ***methyl*** group, the monomer would be ***propylene*** and the polymer, polypropylene.

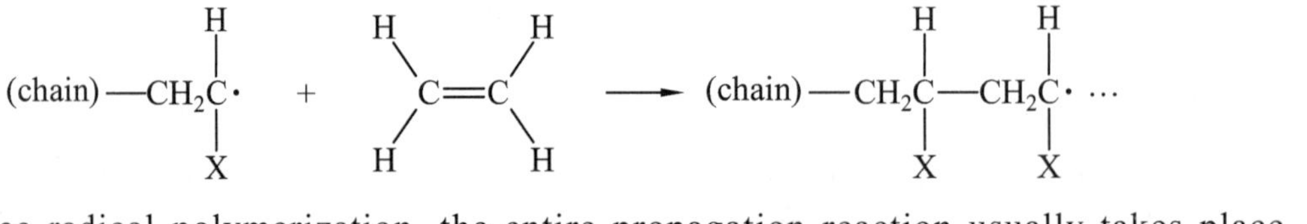

$$\text{(chain)}-CH_2\overset{\overset{\displaystyle H}{|}}{\underset{\underset{\displaystyle X}{|}}{C}}\cdot \; + \; \begin{matrix} H & & H \\ & C=C & \\ H & & H \end{matrix} \longrightarrow \text{(chain)}-CH_2\overset{\overset{\displaystyle H}{|}}{\underset{\underset{\displaystyle X}{|}}{C}}-CH_2\overset{\overset{\displaystyle H}{|}}{\underset{\underset{\displaystyle X}{|}}{C}}\cdot \cdots$$

In free radical polymerization, the entire propagation reaction usually takes place within a fraction of a second. Thousands of monomers are added to the chain within this time. The entire process stops when the termination reaction occurs.

Termination Reaction

In theory, the propagation reaction could continue until the supply of monomers is exhausted. However, this outcome is very unlikely. Most often the growth of a polymer chain is halted by the termination reaction. Termination typically occurs in two ways: ***combination*** and ***disproportionate***.

Combination occurs when the polymer's growth is stopped by free electrons from two growing chains that join and form a single chain. The following diagram depicts combination, with the symbol (R) representing the rest of the chain.

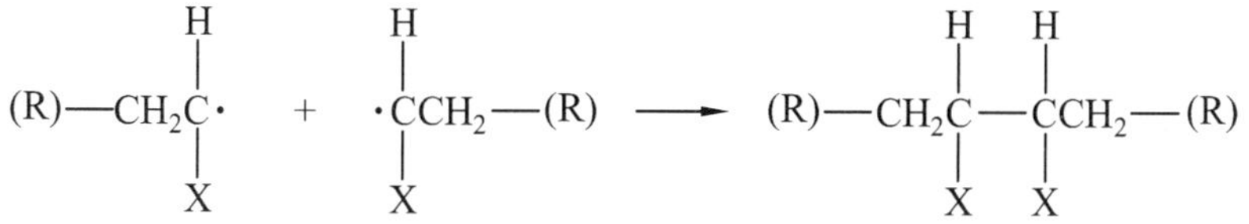

$$\begin{array}{c}\text{H}\\ | \\ \text{(R)—CH}_2\text{C}\cdot\\ | \\ \text{X}\end{array} + \begin{array}{c}\text{H}\\ | \\ \cdot\text{CCH}_2\text{—(R)}\\ | \\ \text{X}\end{array} \longrightarrow \begin{array}{c}\text{H}\quad\text{H}\\ |\quad\ | \\ \text{(R)—CH}_2\text{C—CCH}_2\text{—(R)}\\ |\quad\ | \\ \text{X}\quad\text{X}\end{array}$$

Disproportionation halts the propagation reaction when a free radical strips a hydrogen atom from an active chain. A carbon-carbon double bond takes the place of the missing hydrogen. Termination by disproportionation is shown in the diagram.

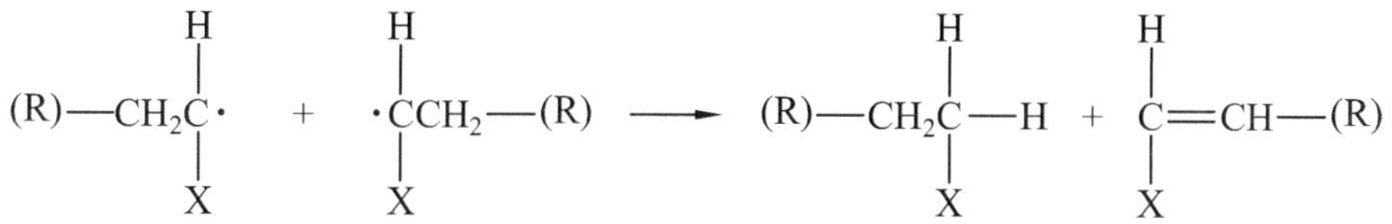

$$\begin{array}{c}\text{H}\\ | \\ \text{(R)—CH}_2\text{C}\cdot\\ | \\ \text{X}\end{array} + \begin{array}{c}\text{H}\\ | \\ \cdot\text{CCH}_2\text{—(R)}\\ | \\ \text{X}\end{array} \longrightarrow \begin{array}{c}\text{H}\\ | \\ \text{(R)—CH}_2\text{C—H}\\ | \\ \text{X}\end{array} + \begin{array}{c}\text{H}\\ | \\ \text{C=CH—(R)}\\ | \\ \text{X}\end{array}$$

Disproportionation can also occur when the radical reacts with an impurity. This is why it is so important that polymerization be carried out under very clean conditions.

Living Polymerization

There exists a type of addition polymerization that does not undergo a termination reaction. This so-called "living polymerization" continues until the monomer supply has been exhausted. When this happens, the free radicals become less active due to interactions with solvent molecules. If more monomers are added to the solution, the polymerization will resume.

Uniform molecular weights (low ***polydispersity***) are characteristic of living polymerization. Because the supply of monomers is controlled, the chain length can be manipulated to serve the needs of a specific application. This assumes that the initiator is 100% efficient.

New Words and Expressions

step-growth polymerization [step grəuθ ˌpɒliməraɪ'zeɪʃn] 逐步聚合（反应）
condensation polymerization [ˌkɔnden'seɪʃən ˌpɔliməraiˈzeiʃən] 缩聚（反应）
chain polymerization [tʃein ˌpɔliməraiˈzeiʃən] 链（式）聚合
addition polymerization [əˈdiʃən ˌpɔliməraiˈzeiʃən] 加聚（反应）
hydroxy [haɪ'drɒksɪ] *adj.* 氢氧根的，羟基的
carboxylic [kɑːbɒk'sɪlɪk] *adj.* 羧基的
kinetics [kɪ'netɪks] *n.* 动力学

polyesterification [pɒlɪesterɪfɪ'keɪʃən] *n.* 聚酯化，酯化缩合反应
polyamide [pɒlɪ'æmaɪd] *n.* 聚酰胺
propagation [ˌprɒpə'geɪʃn] *n.* 增长
free radical polymerization [friː 'rædikəl ˌpɔlimərai'zeiʃən] 自由基聚合
benzoyl peroxide ['benzəuil pə'rɔkˌsaɪd] 过氧化苯甲酰
substituent group [sʌb'stitjuənt gruːp] 取代基
methyl ['meθɪl] *n.* 甲基
propylene ['prəupəliːn] *n.* 丙烯
combination [ˌkɒmbɪ'neɪʃn] *n.* 偶合（终止）
disproportionate [ˌdɪsprə'pɔːʃənət] *n.* 歧化（终止）
polydispersity [pɒlɪdɪs'pɜːsətɪ] *n.* 多分散性

Reading Material

Nomenclature of Polymers

Polymer ***nomenclature*** leaves much to be desired. A standard nomenclature system based on chemical structure as is used for small inorganic and organic compounds is most desired. Unfortunately, the naming of polymers has not proceeded in a systematic manner until relatively late in the development of polymer science. It is not at all unusual of a polymer to have several names because of the use of different nomenclature systems. The nomenclature systems that have been used are based on either the structure of the polymer or the source of the polymer [i.e. , the monomer(s) used in its synthesis] or trade names. Not only have there several different nomenclature systems, but their applications has not always been rigorous. An important step toward standardization was initiated in the 1970s by the ***International Union of Pure and Applied Chemistry*** (***IUPAC***).

Nomenclature Based on Source

The most simple and commonly used nomenclature system is probably that based on the source of the polymer. This system is applicable primarily to polymers synthesized from a single monomer as in addition and ring-opening polymerizations. Such polymers are named by adding the name of the monomer onto the prefix “poly” without a space or hyphen. Thus the polymers from ethylene and ***acetaldehyde*** are named polyethylene and polyacetaldehyde, respectively. When the monomer has a substituted parent name or a multiworded name or an abnormally long name, parentheses are placed around its name following the prefix "poly". The polymers from ***3-methyl-1-pentene,*** vinyl chloride, ***propylene oxide***, ***chlorotrifluoroethylene***, and ***ε-caprolactam*** are named poly (3-methyl-1-pentene), poly (vinyl chloride), poly (propylene oxide), poly (chlorotrifluoroethylene) and poly (ε-caprolactam), respectively. The parentheses are frequently omitted in common usage when naming polymers. Although this will often not present a problem, it is incorrect and in some cases the omission can lead to uncertainty as to the structure of the polymer named.

Nomenclature Based on Structure

A number of the more common condensation polymers synthesized from two different monomers have been named by a semisystematic, structure-based nomenclature system other than the more recent IUPAC system. The name of the polymer is obtained by following the prefix poly without a space or hyphen with parentheses enclosing the name of the structural grouping attached to the parent compound. The parent compound is the particular member of the class of the polymer—the particular ester, amide, urethane, and so on. Thus the polymer from ***hexamethylene diamine*** and ***sebacic acid*** is considered as the substituted amide derivative of the compound sebacic acid, $HO_2C(CH_2)_8CO_2H$, and is named poly (hexamethylene sebacamide). ***Poly (ethylene terephthalate)*** is the polymer from ethylene glycol and terephthalic acid, p-HO_2— C_6H_4—CO_2H. the polymer from trimethylene glycol and ethylene ***diisocyanate*** is poly (trimethylene ethylene-urethane).

A suggestion was made to name condensation polymers synthesized from two different monomers by following the prefix poly with parentheses enclosing the names of the two reactants, with the names of the reactants separated by the term —CO—.

IUPAC Structure-based Nomenclature System

The inadequacy of the preceding nomenclature systems was apparent as the polymer structures being synthesized became increasingly complex. The IUPAC rules allow one to name single-strand organic polymers in a systematic manner based on polymer structure. Single-strand organic polymers have any pair of adjacent repeat units interconnected through only one atom. All the polymers discussed to this point and the large majority of polymers to be considered in this text are single-strand polymers. Double-strand polymers have uninterrupted sequences of rings. A ladder polymer is a double-strand polymer in which adjacent rings have two or more atoms in common.

The basis of IUPAC polymer nomenclature system is the selection of preferred constitutional repeating unit (abbreviated as CRU). The CRU is also referred to as the structural repeating unit. The CRU is the smallest possible repeating unit of the polymer. It is a bivalent unit for a single-strand polymer. The name of the polymer is the name of the CRU in parentheses or brackets prefixed by poly. The CRU is synonymous with the repeating unit except when the repeating unit consists of two symmetric halves, as in the polymers $\left[CH_2CH_2 \right]_n$ and $\left[CF_2CF_2 \right]_n$. the CRU is CH_2 and CF_2, respectively, for polyethylene and ***polytetrafluoroethylene***, while the repeating unit is CH_2CH_2 and CF_2CF_2, respectively.

The constitutional repeating unit is named as much as possible according to the IUPAC nomenclature rules for small organic compounds. The IUPAC rules for naming single-strand polymers dictate the choice of a single CRU so as to yield a unique name, by specifying both the seniority among the atoms or subunits making up the CRU and the direction to proceed along the polymer chain to the end of the CRU. A CRU is composed of two or more subunits when it cannot be named as a single unit.

Trade Names

Special terminology based on trade names has been employed for some polymers. Although trade names should be avoided, one must be familiar with those that are firmly established and commonly used. An example of trade-name nomenclature is the use of the name nylon for the polyamides from unsubstituted, nonbranched ***aliphatic*** monomers. Two numbers are added onto the word "nylon" with the first number indicating the number of methylene groups in the diamine portion of the polyamide and the second number the number of carbon atoms in the diacyl portion. Thus ***poly*(*hexamethylene adipamide*)** and ***poly*(*hexamethylene sebacamide*)** are nylon 6,6 and nylon 6,10, respectively. Variants of these names are frequently employed. The literature contains such variations of the nylon 6, 6 as nylon 66, 66 nylon, nylon 6/6, 6, 6 nylon, and 6-6 nylon. Polyamides from single monomers are denoted by a single number to denote the number of carbon atoms in the repeating unit. Poly(ε-caprolactam) or ***poly*(*6-aminocaproic acid*)** is nylon 6.

In far too many instances trade-name polymer nomenclature conveys very little meaning regarding the structure of a polymer. Many condensation polymers, in fact, seem not to have names. Thus the polymer obtained by the step polymerization of formaldehyde and phenol is variously referred to a phenol-formaldehyde polymer, phenol-formaldehyde resin, phenolic, and phenolic resin, and phenoplast. Polymers of formaldehyde or other aldehydes with urea or melamine are generally referred to as amino resins or aminoplasts without any more specific names. It is often extremely difficult to determine which aldehyde and which amino monomers have been used to synthesize a particular polymer being referred to as an amino resin. More specific nomenclature, if it can be called that, is afforded by indicating the two reactants as in names such as urea-formaldehyde resin or melamine-formaldehyde resin.

A similar situation exists with the naming of many other polymers. Thus the following polymers is usually referred to as "the ***polycarbonate*** from ***bisphenol A***" or polycarbonate. The IUPAC name for this polymer is poly (oxycarbonyloxy-1,4-phenylenedimethylmethylene-1,4- -phenylene).

$$\left[O—CO—O—C_6H_4—C(CH_3)_2—C_6H_4 \right]_n$$

New Words and Expressions

nomenclature [nə'menklətʃə(r)] *n.* 系统命名法

International Union of Pure and Applied Chemistry (IUPAC) 国际纯粹（理论）和应用化学联合会

acetaldehyde [ˌæsɪ'tældəhaɪd] *n.* 乙醛

3-methyl-1-pentene 3-甲基-1-戊烯

propylene oxide ['prəupili:n 'ɔksaid] 环氧丙烷

chlorotrifluoroethylene [klərəʊtraɪflʊə'rəʊɪθɪl] *n.* 三氟氯乙烯

caprolactam [kæprəʊ'læktəm] *n.* 己内酰胺

hexamethylene diamine [ˌheksə'meθili:n dai'æmi:n] 六亚甲基二胺，己二胺

sebacic acid [si'bæsik 'æsid] 葵二酸

poly(ethylene terephthalate) *n.* 聚对苯二甲酸乙二醇酯

diisocyanate [daɪaɪsə'saɪəneɪt] *n.* 二异氰酸盐（酯）

polytetrafluoroethylene [pɒlɪtetrəflʊərə'eθɪli:n] *n.* 聚四氟乙烯

aliphatic [ˌælə'fætɪk] *adj.* 脂肪族的

poly(hexamethylene adipamide) 聚己二酰己二胺

poly(hexamethylene sebacamide) 聚癸二酰己二胺

poly(6-aminocaproic acid) 聚 6-氨基己酸

polycarbonate [ˌpɒli'kɑ:bənət] *n.* 聚碳酸酯

bisphenol A 双酚 A

Lesson 13 Polymer Structure

Although the fundamental property of bulk polymers is the degree of polymerization, the physical structure of the chain is also an important factor that determines the macroscopic properties.

The terms ***configuration*** and ***conformation*** are used to describe the geometric structure of a polymer and are often confused. Configuration refers to the order that is determined by chemical bonds. The configuration of a polymer cannot be altered unless chemical bonds are broken and reformed. Conformation refers to order that arises from the rotation of molecules about the single bonds. These two structures are studied below.

Configuration

The two types of polymer configurations are *cis* and *trans*. These structures can not be changed by physical means (e.g. rotation). The *cis* configuration arises when substituent groups are on the same side of a carbon-carbon double bond. *Trans* refers to the substituents on opposite sides of the double bond.

H H —H_2C H

C=C C=C

—H_2C CH_2— H CH_2—

cis *trans*

Stereoregularity is the term used to describe the configuration of polymer chains. Three distinct structures can be obtained. ***Isotactic*** is an arrangement where all substituents are on the same side of the polymer chain. A ***syndiotactic*** polymer chain is composed of alternating groups and ***atactic*** is a random combination of the groups. Figure 4.2 shows two of the three ***stereoisomers*** of polymer chain.

(a) Isotactic (b) Syndiotactic

Figure 4.2 Two of the three stereoisomers of polymer chain

Conformation

If two atoms are joined by a single by bond then rotation about that bond is possible since, unlike a double bond, it does not require breaking the bond.

The ability of an atom to rotate this way relative to the atoms which it joins is known as an adjustment of the ***torsional angle***. If the two atoms have other atoms or groups attached to them, then configurations which vary in torsional angle are known as conformations. Since different conformations represent varying distances between the atoms or groups rotating about the bond, and these distances determine the amount and type of interaction between adjacent atoms or groups, different conformation may represent different potential energies of the molecule. There are several possible generalized conformations: anti (*trans*), eclipsed (*cis*), and gauche (+ or –).

Other Chain Structures

The geometric arrangement of the bonds is not the only way the structure of a polymer can vary. A branched polymer is formed when there are "side chains" attached to a main chain. A simple example of a branched polymer is shown in Figure 4.3.

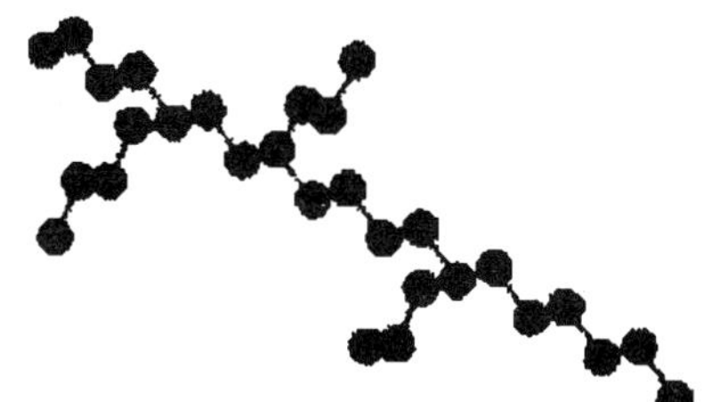

Figure 4.3 Branched polymer

There are, however, many ways a ***branched polymer*** can be arranged. One of these types is called "***star-branching***". Star-branching results when a polymerization starts with a single monomer and has branches radially outward from this point. Polymers with a high degree of branching are called ***dendrimers***. Often in these molecules, branches themselves have branches. This tends to give the molecule an overall spherical shape in three dimensions.

A separate kind of chain structure arises when more that one type of monomer is involved in the synthesis reaction. These polymers that incorporate more than one kind of monomer into their chain are called copolymers. There are three important types of copolymers. A ***random copolymer*** contains a random arrangement of the multiple monomers. A block copolymer contains blocks of monomers of the same type. Finally, a ***graft copolymer*** contains a main chain polymer consisting of one type of monomer with branches made up of other monomers. Figure 4.4, 4.5 and 4.6 displays the different types of copolymers.

Figure 4.4 Block copolymer   Figure 4.5 Graft copolymer   Figure 4.6 Random copolymer

An example of a common copolymer is Nylon. Nylon is an ***alternating copolymer*** with 2 monomers, a 6 carbon diacid and a 6 carbon diamine. The following picture shows one monomer of the diacid combined with one monomer of the diamine:

$$-\overset{\overset{\displaystyle O}{\|}}{C}-(CH_2)_4-\overset{\overset{\displaystyle O}{\|}}{C}-NH-(CH_2)_6-NH-$$

|←— Diacid —→|←— Diamine —→|

Cross-linking

In addition to the bonds which hold monomers together in a polymer chain, many polymers form bonds between neighboring chains. These bonds can be formed directly between the neighboring chains, or two chains may bond to a third common molecule. Though not as strong or rigid as the bonds within the chain, these cross-links have an important effect on the polymer. Polymers with a high enough degree of cross-linking have "memory". When the polymer is stretched, the cross-links prevent the individual chains from sliding past each other. The chains may straighten out, but once the stress is removed they return to their original position and the object returns to its original shape.

One example of cross-linking is vulcanization. In vulcanization, a series of cross-links are introduced into an elastomer to give it strength. This technique is commonly used to strengthen rubber.

New Words and Expressions

configuration [kən,fɪgə'reɪʃn] *n.* 构型
conformation [ˌkɒnfɔː'meɪʃn] *n.* 构象
stereoregularity [stɪərɪəregjʊ'lærɪtɪ] *n.* 立构规整性
isotactic [aɪsəʊ'tæktɪk] *adj.* 全同立构的
syndiotactic [sɪndɪəʊ'tæktɪk] *adj.* 间同立构的
atactic [eɪ'tæktɪk] *adj.* 无规立构的
stereoisomer [ˌsterɪəʊ'aɪsəmə] *n.* 立体异构体
torsional angle 扭转角
branched polymer [brɑːntʃt 'pɒlɪmə(r)] 支化聚合物
star-branching [stɑː(r) 'brɑːntʃɪŋ] 星型支化
dendrimer ['dendrɪmər] *n.* 树形聚合物
random copolymer ['rændəm kəu'pɔlimə] 无规共聚物
graft copolymer [grɑːft kəu'pɔlimə] 接枝共聚物
alternating copolymer ['ɔːltɜːˌnɪtɪŋ kəu'pɔlimə] 交替共聚物

Reading Material

Properties and Applications of Polymers in Relation to Their Molecular Structures

Polymers are the principal light and flexible materials in nature and human society. Compared with most inorganic structural materials, they are of low density ($\approx$1 g·cm^{-3}). They are also the only type of homogeneous material that displays macroscopic, reversible elasticity.

The flexibility and elasticity of polymers is closely related to the rotational degrees of freedom that exist about the covalent skeletal bonds constituting the macromolecular chains and to the lengths of those chains. The net effect of many individual rotations along polymer chains is that those chains can move spatially in response to an external force. In addition to long, flexible macromolecular chains, reversible elasticity requires junction points at the ends of the chains to give an equilibrium network structure defining the spatial positions to which the junction points return after removal of the force. The junction points may be formed by interactions, entanglements, or chemical bonding between chains. Further, in the bulk, liquid state and in solution, polymers display both viscous and elastic properties that relate to the flexibility of their constituent macromolecules and the transient networks formed by the mutual interactions and entanglements of those macromolecules. A well-known example of a type of polymeric viscoelastic liquid is non-drip (***thixotropic***) paint.

The flexibility of polymers not withstanding, many polymer materials are designed to be rigid at their temperatures of use. By using stiffer (e.g., ***aromatic***) chain structures, molecular flexibility can be suppressed. The glass-rubber transition temperature, T_g, is the important characteristic temperature with regard to suppression of flexibility. Below T_g, rotations about the skeletal bonds in a polymer chain may be neglected as significant degrees of freedom. The skeletal bonds then merely librate about equilibrium positions, and significant spatial movements of molecules cannot occur, resulting in rigid (glassy) materials. Engineering plastics, such as aromatic polycarbonate and ***polysulfone***, show values T_g above 150 °C. Additionally, aromatic polyimides having long-term thermal stability above 300 °C have been used in space applications and also in the microelectronics industry as solder-resistant insulation materials.

The flexibility of polymers is also reduced if the chains carry a significant number of strongly interacting groups. Interactions between groups on different molecules enable ordered and, often, crystalline regions to form. The alignment of polymer chains and, hence, the degree of crystallinity can also be enhanced by extending or drawing the molten polymer during processing, as is done, for example, during fiber and, sometimes, film manufacture. Examples are polyamides, PET, and liquid-crystalline polyesters. Because the constituent macromolecules of amorphous polymer materials can never completely disentangle to become mutually aligned, completely crystalline polymers cannot be obtained. However, close control of the physical properties of polymers can be achieved by controlling the formation and concentration of ordered regions.

Network polymers can be made more rigid by increasing the concentration of junction points between normally flexible chains. For example, the vulcanization of natural rubber with about 1% sulfur produces flexible elastomers, and about 30% sulfur forms ***ebonite***, a hard structural material.

An important, unique property of a flexible macromolecular network is that it cannot dissolve in liquids that are solvents for linear polymer chains of the same chemical structure as those that constitute the network. Because a network is essentially a single macromolecule of macroscopic extent, it can only swell in such solvents to form a gel, a soft solid whose structure is maintained by the junction points of the network holding together the polymer chains extended between them. Extremely large increases in volume, up to typically 1000%, can occur. Polymer gels have many important applications, often associated with biology and medicine, for example, as soft contact lenses and super absorbent swabs. In addition to such applications, Merrifield (Nobel Prize, 1984) proposed the use of cross-linked polymer gels in bead form for solid-state syntheses of proteins, facilitating rapid and important developments in biological science and technology over the last two decades. Significant developments of functional polymers have also occurred over the last two decades.

For example, although organic materials in general are insulators, electric conductivity was introduced into polymer films by Shirakawa, Heager, and MacDiarmid (Nobel Prize, 2000) by chemically doping polymers having conjugated backbones. Further significant developments have involved the introduction of photosensitive groups on polymer chains to bring about ***isomerization***, cross-linking, ionization, ***chain cleavage***, or other chemical reactions of polymers under ***photoirradiation***. Such reactions are key processes in microlithography and ***holographic recording*** using ***photoresists*** and photosensitive polymers.

New Words and Expressions

thixotropic [θɪksə'trɒpɪk] *adj.* 触变的

aromatic [ˌærə'mætɪk] *adj.* 芳香的，芳香族的

polysulfone [pɒli:sʌlf'wʌn] *n.* 聚砜

ebonite ['ebənˌaɪt] *n.* 硬橡胶，硬化橡皮，硬橡皮

isomerization [aɪsɒmərai'zeɪʃən] *n.* 异构化

chain cleavage [tʃeɪn 'kli:vɪdʒ] 链断裂

photoirradiation 光致辐照

holographic recording [ˌhɒlə'græfɪk rɪ'kɔ:dɪŋ] 全息图像记录

photoresist [fəʊtəʊrɪ'zɪst] *n.* 光致抗蚀剂，光阻剂

UNIT 5 COMPOSITES

Lesson 14 Introduction to Composites

Composite Definition

A composite is a combined material created by the synthetic assembly of two or more components a selected filler or reinforcing agent and a compatible matrix binder (i.e., a resin) in order to obtain specific characteristics and properties. The components of a composite do not dissolve or otherwise merge completely into each other, but nevertheless do act in concert. The components as well as the interface between them can usually be physically identified, and it is the behavior and properties of the interface that generally control the properties of the composite. The properties of a composite cannot be achieved by any of the components acting alone.

Composite Constituents

The constituents in a composite retain their identity such that they can be physically identified and they exhibit an interface between one another. This concept is graphically summarized in Figure 5.1. The body constituent gives the composite its bulk form, and it is called the matrix. The other component is a structural constituent, sometimes called the reinforcement, which determines the internal structure of the composite. Though the structural component in Figure 5.1 is a fiber, there are other geometries that the structural component can take on, as we will discuss in a subsequent section. The region between the body and structural constituents is called the ***interphase***. It is quite common (even in the technical literature), but incorrect, to use the term interface to describe this region. An interface is a two-dimensional construction—an area having a common boundary between the constituents—whereas an interphase is a three-dimensional phase between the constituents and, as such, has its own properties. It turns out that these interphase properties play a very important role in determining the ultimate properties of the bulk composite. For instance, the interphase is where mechanical stresses are transferred between the matrix and the reinforcement. The interphase is also critical to the long-term stability of a composite. It will be assumed that there is always an interphase present in a composite, even though it may have a thickness of only an atomic dimension.

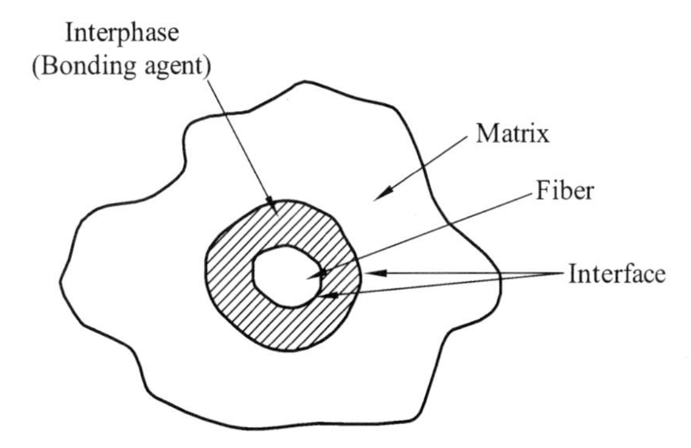

Figure 5.1 Schematic illustration of composite constituents

The chemical composition of the composite constituents and the interphase is not limited to any particular material class. There are metal-***matrix***, ceramic-matrix, and polymer-matrix composites, all of which find industrially relevant applications. Similarly, reinforcements in important commercial composites are made of such materials as steel, E-glass, and Kevlar. Many times a bonding agent is added to the fibers prior to compounding to create an interphase of a specified chemistry.

Composites Classification

There are many ways to classify composites, including schemes based upon:

(1) materials combinations, such as metal-matrix, or glass-fiber-reinforced composites;

(2) bulk form characteristics, such as laminar composites or matrix composites;

(3) distribution of constituents, such as continuous or discontinuous;

(4) function, like structural or electrical composites.

Scheme (2) is the most general, so we will utilize it here.

As shown in Figure 5.2, there are five general types of composites when categorized by bulk form. Fiber composites consist of fibers; with or without a matrix. By definition, a fiber is a particle longer than 100 μm with a length-to-diameter ratio(aspect ratio) greater than 10 : 1. Flake composites consist of flakes, with or without a matrix. A flake is a flat, plate-like material. Particulate composites can also have either a matrix or no matrix along with the particulate reinforcement. Particulates are roughly spherical in shape in comparison to fibers or flakes. In a filled composite, the reinforcement, which may be a three-dimensional fibrous or porous structure, is continuous and often considered the primary phase, with a second material added through such processes as ***chemical vapor infiltration (CVI)***. Finally, laminar composites are composed of distinct layers. The layers may be of different materials, or the same material with different orientation. There are many variations on these classifications, of course, and we will see that the components in fiber, flake, and particulate composites need not be distributed uniformly and may even be arranged with specific orientations.

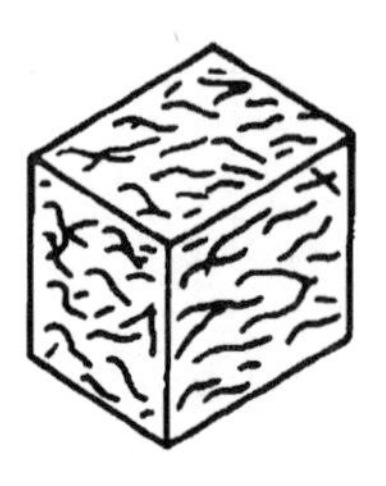

Fiber composite

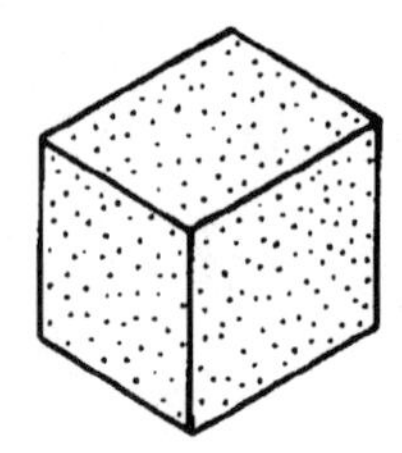

Particulate composite

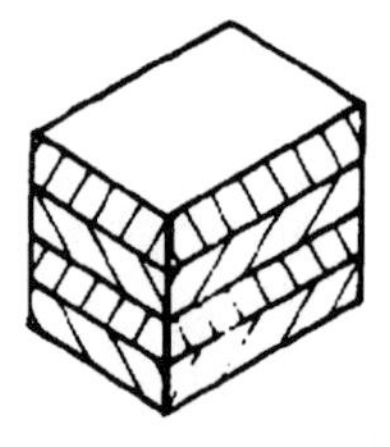

Laminar composite

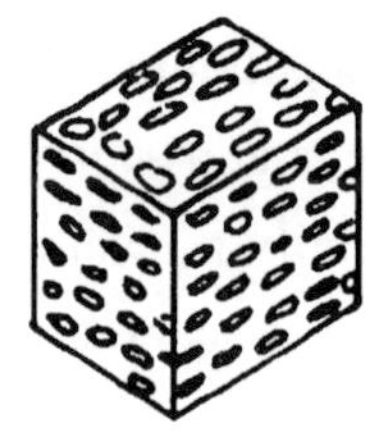

Flake composite

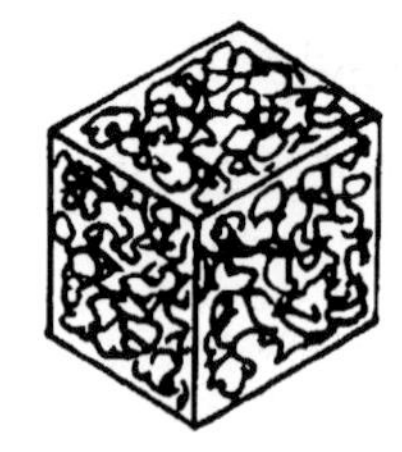

Filled composite

Figure 5.2 Classes of composites

New Words and Expressions

interphase ['ɪntəfeɪz] *n.* 界面相

matrix ['meɪtrɪks] *n.* 复合材料中的基体

chemical vapor infiltration (CVI) ['kemikəl 'veipə ˌinfil'treiʃən] 化学气相渗透

Reading Material

Properties of Composites

We will consider the results of incorporating fibers in a matrix. The matrix, besides holding the fibers together, has the important function of transferring the applied load to the fibers. It is of great importance to be able to predict the properties of a composite, given the component properties and their geometric arrangement.

Isotropic vs. Anisotropic

Fiber reinforced composite materials typically exhibit anisotropy. That is, some properties vary depending upon which geometric axis or plane they are measured along.

For a composite to be ***isotropic*** in a specific property, such as CTE (the coefficient of thermal expansion) or Young's modulus, all reinforcing elements, whether fibers or particles, have to be randomly oriented. This is not easily achieved for discontinuous fibers, since most processing methods tend to impart a certain orientation to the fibers. One example is the classic Sink-Core-Skin pattern seen for injection molded short fiber composites.

Continuous fibers in the form of sheets are usually used to deliberately make the composite anisotropic in a particular direction that is know to be the principally loaded axis or plane.

Rule of Mixtures

The Rule of Mixtures is a rough tool that considers the composite properties as volume weighted averages of the component properties. It is important to realize that this rule works accurately only in certain simple situations, such as determining composite density and elastic modulus. For most other properties, this provides only a rough estimate for initial design purposes. Below are some equations derived for unidirectional continuous fiber composites. The derivations can be found in any introductory book on composite mechanics. The principle used is that in ***longitudinal*** direction, both fibers and matrix have the same strain (***isostrain***) and in ***transverse*** direction, both fibers and matrix have the same stress (***isostress***).

The subscripts f, m, v and c refer to fiber , matrix, voids and composite respectively.

The subscripts l and t refer to longitudinal and transverse respectively.

E is the Young's modulus, and v (rho) is density. α (alpha) is the coefficient of thermal expansion (CTE).

Lowercase v refers to volume, whereas uppercase V refers to volume fraction (volume of a component divided by total volume).

a. Density

$$\rho_c = \frac{m_c}{v_c} = \frac{m_f + m_m}{v_f + v_m + v_v} = \frac{\rho_f v_f + \rho_m v_m}{v_f + v_m + v_v} = \frac{\rho_f V_f + \rho_m V_m}{l}$$

b. Young's modulus

Longitudinal

$$E_{\mathrm{cl}} = E_{\mathrm{f}} V_{\mathrm{f}} + E_{\mathrm{m}} V_{\mathrm{m}}$$

Transverse

$$\frac{1}{E_{\mathrm{ct}}} = \frac{V_{\mathrm{m}}}{E_{\mathrm{m}}} + \frac{V_{\mathrm{f}}}{E_{\mathrm{f}}}$$

c. Coefficient of Thermal Expansion

Longitudinal

$$\alpha_{\mathrm{cl}} = \frac{\alpha_{\mathrm{m}} E_{\mathrm{m}} V_{\mathrm{m}} + \alpha_{\mathrm{f}} E_{\mathrm{f}} V_{\mathrm{f}}}{E_{\mathrm{m}} V_{\mathrm{m}} + E_{\mathrm{f}} V_{\mathrm{f}}}$$

Transverse

$$\alpha_{\mathrm{ct}} = (1 + \nu_{\mathrm{m}}) \alpha_{\mathrm{m}} V_{\mathrm{m}} + \alpha_{\mathrm{f}} V_{\mathrm{f}}$$

CTE Mismatch

Because of a difference in the thermal expansion properties of fibers and matrix, the composite is not allowed to deform uniformly under thermal stress, and this can lead to microcracking of the matrix and debonding at the interface. This is a particularly important concern in dental composite materials where thermal stresses are significant.

Interface, Fracture Propagation and Toughness

In a ductile matrix, like most polymers and metals, a strong interfacial bond is important, since the fibers carry most of the load in such matrices. Fibers tend to fail first, usually by cohesive failure through the fiber cross-section. This is because the fibers cannot strain as much as the matrix (e.g. carbon in epoxy). Cracks are few, and tend to propagate slowly. When the cracks hit the interface, strong interfacial bonds stop them.

In a brittle matrix, like ceramics, the matrix carries most of the load, which is usually compressive (like in teeth or bone), and fibers are added only to increase toughness. That is, to increase the time to catastrophic failure by holding the matrix together after cracking. Fibers here are more ductile than the matrix (e.g. glass in alumina) and the matrix fails first. As the cracks propagate and reach the interface, a weak interfacial bond is desired. This enhances debonding, and the cracks are not stopped, but deflected along the length of the fibers. This effectively delays the time it takes the cracks to propagate through the entire matrix, and thus increases toughness.

New Words and Expressions

isotropic [ˌaɪsə'trɒpɪk] *adj.* 各向同性的
longitudinal [ˌlɒŋgɪ'tjuːdɪnl] *adj.* 经度的，纵向的
isostrain 等应变
transverse ['trænzvɜːs] *adj.* 横向的，横断的
isostress 等应力
mismatch ['mɪsmætʃ] *v.* 不匹配

Lesson 15 Polymeric Composite Materials

Introduction

Principal advantage of composite materials resides in the possibility of combining physical properties of the constituents to obtain new structural or functional properties. Composite materials appeared very early in human technology, the "structural" properties of straw were combined with a clay matrix to produce the first construction material and, more recently, steel reinforcement opened the way to the ***ferroconcrete*** that is the last century dominant material in ***civil engineering***. As a matter of fact, the modern development of polymeric materials and high modulus fibres (carbon) introduced a new generation of composites. The most relevant benefit has been the possibility of energetically convenient manufacturing associated with the low weight features. Due to the possibility of designing properties, composite materials have been widely used, in the recent past, when stiffness/weight, strength/weight, ability to tailor structural performances and thermal expansion, corrosion resistance and fatigue resistance are required. Polymeric composites were mainly developed for aerospace applications where the reduction of the weight was the principal objective, irrespective of the cost. The scientific efforts in this field such as automotive, naval transportation and civil engineering but the high cost still limits their applications. A continuous task has been making composite components economically attractive. The effort to produce economic attractive composite components has resulted in several innovative manufacturing techniques currently being used in the composite industry.

State of the Art

Nowadays, technology is devoted to the development of new materials able to satisfy specific requirements in terms of both structural and functional performances.

The need of exploring new markets in the field of polymeric composites has recently driven the research in Europe towards the development of new products and technologies. In particular, activities on thermoplastic based composites and on composites based on natural occurring materials (***environmentally friendly***, ***biodegradable systems***) have been of relevant interest in many European countries.

Since the beginning of the 1990s, U.S. and Japan have recognized the need of expanding composite applications. In the field of materials, Japan put more emphasis than U.S. on thermoplastic and high temperature resins. Moreover, due to the large extent of the textile industry in Japan, textile preforming is significantly more advanced than in U.S., and this could lead to the development of cost-efficient automated computer-controlled looms for complex textile shapes. In contrast to the U.S. approach of developing computational models to better understand manufacturing processes, Japanese manufacturing science appears to reside in experienced workers who develop understanding of the process over long period of time. However, Japanese process and product development methods are based on concurrent engineering ***methodology***, which is based on the integration of product and process design.

Biomedical is another important field where composites are applied. Materials, able to simulate the complex structural properties of the natural tissues, which are composite in nature, have been developed but there are still few applications. This is due to the delay in the technology transfer from different areas (composite industry and biomedical) and to the lack of cross-disciplinary strategies. In this field, U.S. maintain a leadership role but major centres exist in EU and Japan. Japan recognized the necessity of establishing a "National Institute for Advanced Interdisciplinary Research" which is devoted to research on subjects combining elements from various fields that cannot be adequately treated within the bounds of traditional divisions of science. Among other topics, soft and hard tissue engineering are considered of relevant interests. Tissue engineering activities are growing, as well and strong R&D programmes are present in EU and Japan, even though U.S. have a leadership role in the field.

Trends

The main trends in the structural composite field are related to the reduction of the cost which cannot only be related to the improvement in the manufacturing technology, but needs an integration between design, material, process, tooling, quality assurance, manufacturing. Moreover, the high-tech industry, such as telecommunication, where specific functional properties are the principal requirements, will take advantages by the composite approach in the next future. The control of the filler size, shape and surface chemical nature has a fundamental role in the development of materials that can be utilized to develop devices, sensors and actuators based on the tailoring of functional properties such as optical, chemical and physical, ***magneto-elastic*** etc. Finally, a future technological challenge will be the development of a new class of smart composite materials whose ***elasto-dynamic response*** can be adapted in real time in order to significantly enhance the performance of structural and mechanical systems under a diverse range of operating conditions.

Over the long period term, U.S. and Japan believe that advances in the materials area would prompt new breakthroughs in the area of composites. In face, the current emphasis is on "fourth-generation" materials, i.e. those that are designed by controlling the behavior of atoms and electrons, and which provide carefully tailored functional gradients.

Conclusions

Research activities, aimed to expand the applications in composite industry, must be addressed to improve manufacturing composite technology, through a better integration of product and process design; to develop new constituent materials with better performances and/or for the tailoring of structural and functional properties for special applications and for the development of new processes and new manufacturing technologies.

Expected breakthroughs are related to the development of multi-component materials with anisotropic and non-linear properties, able to impart unique structural and functional properties. Applications include smart systems, able to recognize and to adapt to external stimuli, as well as anisotropic and active composite systems to be used as scaffold for tissue engineering and other biomedical applications.

New Words and Expressions

ferroconcrete ['ferəʊ'kɒnkriːt] *n.* 钢筋混凝土

civil engineering ['sivl ˌendʒi'niəriŋ] 土木工程

environmentally friendly [enˌvaɪrən'mentlɪ 'frendli] 环境友好的

biodegradable system [ˌbaɪəʊdɪ'greɪdəbl 'sɪstəm] 生物降解系统

methodology [ˌmeθə'dɒlədʒi] *n.* 方法学

magneto-elastic [mæg'niːtəʊ ɪ'læstɪk] *adj.* 磁弹性的

elasto-dynamic response 弹性动力响应

Reading Material

Ceramic Matrix Composites

Introduction

Ceramic matrix composites (CMCs) have been developed to overcome the intrinsic brittleness and lack of reliability of ***monolithic*** ceramics, with a view to introduce ceramics in structural parts used in severe environments, such as rocket and jet engines, gas turbines for power plants, heat shields for space vehicles, fusion reactor first wall, aircraft brakes, heat treatment furnaces, etc. It is generally admitted that the use of CMCs in advanced engines will allow an increase of the temperature at which the engine can be operated and eventually the elimination of the cooling fluids, both resulting in an increase of yield. Further, the use of light CMCs in place of heavy ***superalloys*** is expected to yield significant weight saving. Although CMCs are promising ***thermostructural*** materials, their applications are still limited by the lack of suitable reinforcements, processing difficulties, sound material data bases, lifetime and cost.

Ceramic Matrix Composite Spectrum

A given ceramic matrix can be reinforced with either discontinuous reinforcements, such as particles, whiskers or chopped fibres, or with continuous fibres. In the first case, the enhancement of the mechanical properties, in terms of failure strength and toughness, is relatively limited but it can be significant enough for specific applications, a well known example being the use of ceramics reinforced with short fibres in the field of the cutting tools (SiC_W/ Si_3N_4 composites). Among the discontinuous reinforcements, whiskers are by far the most attractive in terms of mechanical properties. Unfortunately, their use raises important health problems both during processing and in service. Conversely, continuous reinforcements, such as fibre yarns, are much more efficient, from a mechanical stand-point, but they are more expensive and more difficult to use in a ceramic matrix in terms of material design and processing.

There is a wide spectrum of CMCs depending on the chemical composition of the matrix and reinforcement.

Non-oxide CMCs are by far those which have been the most studied. Such a choice could appear surprising since the atmosphere in service is often oxidizing. That choice could be explained as follows. The most performant fibres, in terms of stiffness, failure strength, refractoriness and density are non-oxide fibres, i.e. carbon and silicon carbide fibres. Further, carbon fibres are extensively used in volume production of ***polymer matrix composites***. As a result, they are much cheaper than all the other fibres (glass fibres excepted). Second, in order to avoid compatibility problems, which are crucial oxide fibres are preferably embedded in non-oxide matrices. Hence, the first non-oxide CMCs have been carbon/carbon (C/C) composites. They have been initially designed and produced for use in rocket engines and reentry heat shields, i.e. under extremely severe service conditions but short lifetimes. In a second step, C/SiC and SiC/SiC composites were developed in order to increase the oxidation resistance of the materials and hence their lifetimes in oxidizing atmospheres. Silicon nitride was also used as matrix although it is less stable at high temperatures than silicon carbide.

Oxide-CMCs would obviously be the best choice, from a thermodynamic standpoint, for long term applications in oxidizing atmospheres. Unfortunately, oxide fibres, although they are refractory, tend to undergo grain growth at high temperatures, (which results in a fibre strength degradation) and exhibit a poor creep resistance. Further, they display much higher densities than say carbon fibres (4 $g \cdot cm^{-3}$ for alumina versus 2 for carbon). Attempts have been made to improve the high temperature properties of oxide fibres with limited success. Despite these disadvantages, Al_2O_3/Al_2O_3 and derived CMCs have been, and are still, extensively studied.

Expected Breakthroughs and Future Visions

The future of CMCs is directly depending on progress that would be achieved in the availability of higher performance constituents (fibres, interphases and tailored matrices) as well as in processing cost reduction.

As far as the reinforcements are concerned, two main breakthroughs are expected: (1) the availability of a low cost non-oxide fibre that could be used up to about 1500 °C and (2) the development of a refractory oxide fibre resistant to grain growth and creep. Oxygen-free quasi-stoichiometric SiC fibres display much better high temperature properties than their Si-C-O counterparts fabricated from poly-carbosilane according to the Yajima's route. However, they are too costly (with respect to carbon fibres and CMC volume production) and their failure strain is too low. Amorphous Si-B-C-N fibres, presented as creep resistant at high temperature, are still at a development stage. Although alumina-based binary oxide fibres, e.g. mullite/alumina or alumina/YAG fibres, represent a significant progress in terms of creep resistance with respect to pure α-alumina fibres, further improvement is still necessary to match the high temperature properties of non-oxide fibres. Finally, nanotubes, with their outstanding mechanical properties, may raise problems similar to those previously encountered with whiskers.

The spectrum of suitable interphase materials that could be used in a realistic manner in CMCs remains extremely narrow. In non-oxide CMCs, there is presently no alternative to the carbon-based interphases. Boron nitride is obviously the only potential candidate. However, its sensitivity to moisture when poorly crystallized and its low bonding to SiC-based fibres are subjects of concern. Solving these two problems will be an interesting breakthrough. The search for new interphase materials, displaying a better oxidation resistance than carbon and boron nitride and which could be easily deposited in situ in multidirectional fibre preforms, should be strongly encouraged.

The recent discovery of the ***self-healing multilayered matrices*** has been an important breakthrough since it permits the use of non-oxide CMCs in oxidizing atmospheres. The concept should obviously be further developed in terms of material selection.

Finally, the processing cost of CMCs should be reduced (although the main contribution to the total cost of a given part is presently, e.g. in a SiC/SiC composite, that of the reinforcement, as previously mentioned). Gas phase route processes with a significant reduction of the overall densification time, liquid phase route processes with a limited number of PIP-sequences (through the use of appropriate precursors), both being compatible with CMC volume production, would obviously be significant breakthroughs.

New Words and Expressions

ceramic matrix composites (CMCs) [sə'ræmɪk 'meɪtrɪks 'kɒmpəzɪts] 陶瓷基复合材料
monolithic [ˌmɒnə'lɪθɪk] *adj.* 庞大而单一的
superalloy ['sju:pə'ælɔɪ] *n.* 超合金，超耐热合金
thermostructural [θɜ:'məʊstrʌktʃərəl] *adj.* 结构热学的
polymer matrix composites ['pɒlɪmə(r) 'meɪtrɪks 'kɒmpəzɪts] 聚合物基复合材料
self-healing ['selfh'i:lɪŋ] 自愈合，自愈性

PART Ⅱ

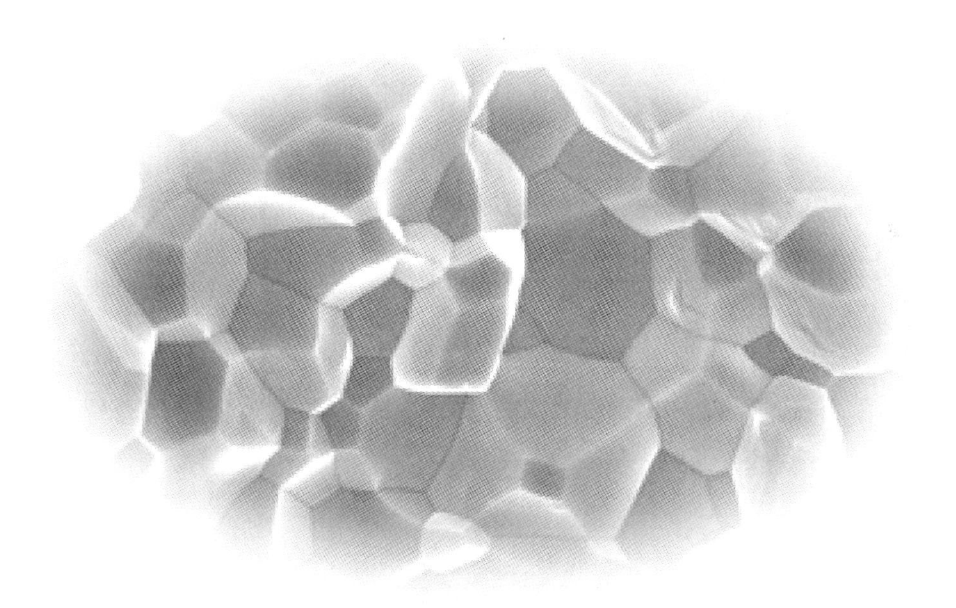

UNIT 6　外文文献检索

Lesson 16　三大检索工具简介

SCI、EI、ISTP 是世界著名的三大科技文献检索系统，其收录论文的情况是评价国家、单位和科研人员的成绩、水平以及进行奖励的重要依据之一。

SCI——Science Citation Index，科学引文索引。

EI——The Engineering Index，工程索引。

ISTP——Index to Scientific and Technical Proceedings，科技会议录索引。

一、SCI

科学引文索引（Science Citation Index，SCI）是一种多学科的科技文献检索工具，由美国科学信息研究所（Institute for Scientific Information，ISI）1961 年创办出版的引文数据库。SCI 以布拉德福（S. C. Bradford）文献离散律理论和加菲尔德（E. Garfield）引文分析理论为基础，结合论文被引用频次的统计数据，对学术期刊和科研成果进行多方位的评价研究，从而评判一个国家或地区、科研单位、个人的科研产出绩效，反映其在国际上的学术水平。因此，SCI 是目前国际上公认并广泛使用的科学引文索引数据库和科技文献检索工具。

1. SCI 收录范围

SCI 收录全世界出版的数学、物理、化学、农业、林业、医学、生命科学、天文、地理、环境、材料、工程技术等自然科学各学科的核心期刊 3700 多种，每年数目略有增减。SCI 文献来源涵盖 45 个国家或地区的核心期刊约 3500 种，SCI-E（SCI Expanded）是 SCI 的扩展版，收录期刊 8400 余种，涵盖各个领域。

2. SCI 结构体系

SCI 由引文索引、来源索引和轮排主题索引三部分组成。引文索引是一种以科技期刊、专利、专题丛书、技术报告等文献资料所发表论文参考文献（引文）的作者、题目、出处等项目，按照引用与被引用的关系进行排列而编制的索引；引文索引又可分作者引文、团体著者、匿名和专利引文索引。来源索引提供编入本期 SCI 的全部来源文献的篇名及其出处，可分为来源出版物一览表、团体索引、来源作者索引等。轮排主题索引由计算机在来源文献的篇名中自动抽取数个具有独立检索意义的关键词，所有关键词实行轮排，用于检索某个领域或课题被 SCI 收录的相关文献。

3. SCI 功能

通过 SCI 统计数据，可以掌握某位著者的论文曾被何人引用，该著者有多少篇论文被引用过，某篇论文被引用次数等信息，以此了解该著者科学研究进展，同时还可以掌握国际上同行或竞争对手的研究现状。通过 SCI 可以了解到世界上有多少人共同研究相同或相近的科研课题，充分掌握该领域的科研动态，通过团体索引可以掌握某一国家、地区、某一学术机构或科研单位的研究进展，了解科学研究现状。

我国 1987 年引入 SCI 等国外著名的检索工具作为期刊科技工作评价的依据。每年由中国科技信息研究所进行年度统计，年末的统计结果公布成为全国各地科研机构、高校、医疗单位等关心的大事，其学术论文被 SCI 引用情况已成为评价学术水平和科研实力的重要标准。

4. SCI 检索方法

SCI 有印刷版（print）、光盘版（CD-ROM）、网络版（web）和联机版（online）四种形式，这里主要介绍 SCI 网络版的检索方法。

SCI 网络版可通过 Web of Science 检索，该系统是一个大型、多学科、核心期刊引文索引数据库，包括 ISI 公司的三大引文数据库：SCI、SSCI（Social Sciences Citation Index，社会科学引文索引）、A&HCI（Arts & Humanities Citation Index，艺术与人文科学引文索引）和两个化学信息事实型数据库：CCR（Current Chemical Reactions）、IC（Index Chemicus）。Web of Science 提供被引文献检索（Cited Reference Search）的检索机制，可以轻松地回溯或追踪学术文献，既可以“越查越旧”，也可以“越查越新”，超越学科与时间的局限，迅速地发现在不同学科、不同年代所有与自己研究课题相关的重要文献。

检索步骤：

（1）进入主网页，选择数据库（Select database(s) and timespan）。在 SCI-E、CCR、IC 三个数据库前的方框上打钩，数据库默认为选择“SCI-E”。

（2）检索。Web of Science 提供 Easy Search（简单检索）和 Full Search（充分检索）两种检索方式。Easy Search 可按主题、人物、作者地址进行检索；Full Search 又包括 General Search（常规检索）和 Cited Reference Search（引文检索）两种，General Search 可通过主题、作者、刊名、作者地址进行检索，Cited Reference Search 可通过被引作者、被引文献进行检索，同时可选择时间范围、限制文献类型及语种，检索结果还可按多种方式排序。

（3）检索结果显示。选择结果排序方式，可按 Latest Date（最新纪录）、Times Cited（被引次数）、Relevance（相关度）、First Author（第一作者姓氏字顺）、Source Title（来源刊名字顺），默认方式按 Latest Date 排序。

5. SCI 期刊影响因子

影响因子（Impact factor，IF）是 ISI 的期刊引证报告（Journal Citation Reports，JCR）中公布的一项数据，是国际上通行的期刊评价指标，同时也是衡量学术期刊影响力的一个重要指标。影响因子是指某期刊前两年（S, T）发表的论文在统计当年（U）的被引用总次数 X 除以该期刊在前两年内发表的论文总数 Y，计算公式为 $IFU = X(S, T)/Y(S, T)$。

二、EI

工程索引（Engineering Index，EI），1884 年由美国工程信息公司（Elsevier Engineering Information Inc.）出版的一种著名检索刊物。EI 选用世界上工程技术类几十个国家和地区 15 个语种的 4500 余种期刊和 2000 余种会议录、科技报告、标准、图书等出版物，每年大约增加文献 20 万篇，它是人们获取工程学及其相关领域信息的最主要来源。

1. EI 收录范围

其收录范围涉及工程及相关学科的各个领域。

2. EI 数据库

Engineering Village 是由美国工程信息公司在因特网上提供的网络数据库，它的核心产品 EI。EI Compendex 数据库从 2009 年 1 月起，所收录的数据不再分核心数据和非核心数据，所有数据都是以 EI Compendex 的形式存在。现在 EI 数据库收录文献只分为两类：期刊检索（Journal Article，JA）和会议检索（Conference Article，CA）。

3. EI 检索方法

（1）进入主网页，选择数据库，进入检索界面。

（2）检索

EI 检索有三种检索方法，字段检索，在某个字段后的检索框内输入检索词，检索词之间可使用逻辑算符和位置算符，点击“Search”按钮，开始检索。

索引检索，点击字段名后的“index”，弹出该字段的索引列表，在“Jump To”后的框内输入检索词，然后点击“Jump To”按钮，可快速定位到该检索词；选择某个具体的检索词，点击“Search”按钮，开始检索。

复合检索，经过两次或两次以上检索后，在检索历史框列出各次检索的步号、该次检索采用的检索策略及其命中结果。选择两次或两次以上检索步骤（必须是在同一个数据库中进行的检索步骤），点击“INTERSECT”按钮，各检索步骤之间以 AND 组合；点击“UNION”按钮，各检索步骤之间以 OR 组合。或选择某一检索步骤，再在某个字段后输入检索词，点击“NARROW”按钮，系统即进行 AND 运算；点击“WIDEN”按钮，进行 OR 运算；点击“EXCLUDE”按钮，进行 NOT 运算。

（3）检索结果显示

在检索历史框中，选择某个检索步骤，点击“DISPLAY”按钮，以全记录格式显示检索结果，使用翻页按钮，浏览其他记录；点击“TITLE LIST”按钮，显示命中文献的标题。如果想一次浏览多篇相关文献，使用“Ctrl-click”选定标题，再点击“DISPLAY”按钮。“SORT”按钮用于排序，可选定要排序的字段，作升序或降序排列。

由于 Dialog 系统没有提供选择、保存、打印检索结果的功能，必须使用浏览器的保存或打印命令。如果要保存多个记录，只能在浏览多篇文献窗口，保存或打印所有记录。

三、ISTP

科技会议录索引(Index to Scientific and Technical Proceedings，ISTP)，也是由美国科学信息所（ISI）编辑出版，1978 年创刊，报道世界上每年召开的科技会议的会议论文。在每年召开的国际重要学术会议中，有 75%～90%的会议被此索引引录，内容涉及科学技术的各个领域。

ISTP 数据库是美国科学情报研究所（ISI）提供的 ISI Proceedings 数据库，ISI Proceedings 包括 ISTP（科学技术会议录索引）和 ISSHP（社会科学及人文科学会议录索引）两大会议录索引集，汇集了世界上最新出版的会议录资料，包括专著、丛书、预印本以及来源于期刊的会议论文，提供了综合全面、多学科的会议论文资料。ISI Proceedings 数据库通过 ISI Web of Knowledge 检索平台使用，检索方法与 Web of Science 类似。

四、其他检索系统

1. ISR

科学评论索引（Index to Scientific Reviews，ISR），是 ISI 公司出版的半年刊，每年收录 200 多种综述出版物和 3000 多种期刊中的综述类文献，学科范围与 SCI 基本相同。同时，ISR 与 SCI、EI、ISTP 通常被认为是世界著名的四大科技文献检索系统。

2. ISI 引文系列

SSCI（社会科学引文索引）、A&HCI（艺术与人文引文索引）是 SCI 的姐妹索引，收录学科范围分别是全球 1700 多种社会科学期刊和 1100 多种人文艺术科学期刊。

3. 中国科学引文索引数据库

简称 CSCI 或 CSCD，中国科学院文献情报中心研制出版，收录我国出版的千余种中、英文重要期刊，专业覆盖数、理、化、农、医及工程技术各领域。数据自 1989 年起每年更新，包括文献收录和被引用情况。

Lesson 17　常用外文电子资源

1. SpringerLink

德国施普林格（Springer-Verlag）是世界上著名的科技出版集团，通过 SpringerLink 系统提供学术期刊及电子图书的在线服务，可检索阅读著名的德国施普林格科技出版集团出版的全文电子期刊，是科研人员的重要信息源。SpringerLink（Springer 线上出版服务平台）中的期刊及图书等所有资源划分为 12 个学科：建筑学、设计和艺术；行为科学；生物医学和生命科学；商业和经济；化学和材料科学；计算机科学；地球和环境科学；工程学；人文、社科和法律；数学和统计学；医学；物理和天文学等，并包含了很多跨学科内容。

2002 年 7 月开始，Springer 公司在我国开通了 SpringerLink 服务，目前 SpringerLink 已迅速成为面向全球科研服务的最大在线全文期刊数据库和丛书数据库之一。

2. Elsevier

荷兰爱思唯尔（Elsevier）出版集团是全球最大的科技与医学文献出版发行商之一，已有 180 多年历史。该公司出版的期刊是国际公认的高水平学术期刊，大多数都被 SCI、EI 所收录，属国际核心期刊。ScienceDirect 系统是 Elsevier 公司的核心产品，自 1999 年开始向读者提供电子出版物全文的在线服务，包括 Elsevier 出版集团所属的 2 200 多种同行评议期刊和 2 000 多种系列丛书、手册及参考书等，涉及数学、物理、化学、天文学、医学、生命科学、商业及经济管理、计算机科学、工程技术、能源科学、环境科学、材料科学等。ScienceDirect OnSite（SDOS）是国内 11 所学术图书馆于 2000 年首批联合订购 SDOS 数据库中 1998 年以来的全文期刊，并已在清华大学图书馆设立镜像站点。

3. EBSCO

该数据库是 EBSCO 公司（美）提供的学术信息、商业信息网络版数据库，包括学术期刊数据库（Academic Search Elit）和商业资源数据库（Business Source Premier）。数据库将二次文献与一次文献捆绑在一起，为最终用户提供文献获取一体化服务，检索结果为文献的目录、文摘、全文（PDF 格式）。

学术期刊数据库包括生物科学、工商经济、咨询科技、通信传播、工程、教育、艺术、医药学等领域的期刊，约 2700 种，其中全文有 1240 种。SCI & SSCI 收录的核心期刊为 993 种（全文有 350 种）。

商业资源数据库包括经济学、经济管理、金融、会计、劳动人事、银行以及国际商务等领域的期刊，约 2000 种，其中全文有 1600 余种。SCI & SSCI 收录的核心期刊为 398 种（全文有 145 种）。

4. Wiley InterScience

John Wiley & Sons Inc.（美）是有 200 年历史的国际知名专业出版机构，在化学、生命科学、医学以及工程技术等领域学术文献的出版方面颇具权威性，Wiley InterScience 是其综合性的网络出版及服务平台，目前使用的界面是 2003 年 8 月推出的最新版本。在该平台上提供全文电子期刊、电子图书和电子参考工具书的服务。

5. PQDD 全文——ProQuest 学位论文全文

PQDD 全称是 ProQuest Digital Dissertations，是世界著名的学位论文数据库，收录欧美 1000 余所大学文、理、工、农、医等领域的博士、硕士学位论文，是学术研究中十分重要的信息资源。

国内若干图书馆、文献收藏单位每年联合购买一定数量的 ProQuest 学位论文全文（PDF 格式），提供网络共享，即：凡参加联合订购成员馆均可共享整个集团订购的全部学位论文资源。目前"ProQuest 学位论文全文中国集团"在国内已建立了三个镜像站，您可登录其中任一个网址检索该数据库，下载硕博士学位论文的 PDF 全文：

（1）ProQuest 学位论文全文数据库（CALIS 镜像站）。

（2）ProQuest 学位论文全文数据库（上海交通大学镜像站）。

（3）ProQuest 学位论文全文数据库（中国科学技术信息研究所镜像站）。

6. Netlibrary（英文电子图书）

NetLibrary 处在美国科罗拉多州波尔德尔市，于 1999 年成立，是世界上向图书馆提供电子图书的主要提供商。NetLibrary 于 2002 年 1 月 25 日成为 OCLC 联机计算机图书馆中心的下属部门，是当前世界上 eBook 的主要提供商。目前，世界上 7000 多个图书馆通过 NetLibrary 存取电子图书，其中包括哥伦比亚大学，斯坦福大学，加州大学伯克莱分校，以及世界上其他成千的大小图书馆。

OCLC NetLibrary 目前提供 400 多家出版社出版的 60 000 多种电子图书，并且每月增加约 2000 种，这些电子图书覆盖所有主题范畴，约 80%书籍是面向大学程度的读者。大多数 NetLibrary 的电子图书内容新颖，近 90%的电子图书是 1990 年后出版的。

7. Kluwer

荷兰 Kluwer Academic Publisher 是具有国际性声誉的学术出版商，它出版的图书、期刊一向品质较高，备受专家和学者的信赖和赞誉。Kluwer Online 是 Kluwer 出版期刊的网络版，专门基于互联网提供 Kluwer 电子期刊的查询、阅览服务。目前通过 CALIS 镜像站，用户可以使用 Kluwer Academic Publisher 的 800 种电子刊，并可以检索、阅览和下载全文，并免付 Internet 网络费。涵盖学科有：材料科学、地球科学、电气电子工程、法学、工程、工商管理、化学、环境科学、计算机与信息科学、教育、经济学、考古学、人文科学、社会科学、生物学、数学、天文学、心理学、医学、艺术、语言学、哲学等 24 个学科。

由于 Kluwer 公司已与德国施普林格（Springer）出版公司合并，目前通过 SpringerLink 系统也可以访问到 Kluwer 出版的期刊，且访问的年限更长，数据质量更高。

8. Blackwell

英国 Blackwell Publishing 是全球最大的学协会出版商，与世界各地 600 多个学协会组织和专业机构合作，是全球三大学术出版社之一，通过 Blackwell Synergy 平台提供电子期刊的在线服务。其中有 100 多种期刊先于印刷刊在网上提供，时效性强，还有对引文和参考文献的链接。

Blackwell 出版的期刊学术质量很高，大多是各学科领域内的核心刊物，在医学、护理学、生态学、农业、生物学、社会科学、经济学、数学、工程学、建筑学和计量学等领域享有盛誉，被 SCI 和 SSCI 收录的期刊占所有期刊总数近 70%。目前 Synergy 中已包含 881 种全文学术期刊。

9. CA

美国《化学文摘》（Chemical Abstracts，CA）是世界上著名的检索刊物之一。由美国化学学会文摘服务社编辑。CA 收录内容主要从世界各国涉及化学化工方面的 8000 余种期刊、专利、科技报告、专著、会议录、学位论文等各类文献中摘录而来。据称，CA 已摘录了世界上有关化学的 98%文献，报道的内容几乎涉及了化学家感兴趣的所有领域，其中除包括无机化学、有机化学、分析化学、物理化学、高分子化学外，还包括冶金学、地球化学、药物学、毒物学、环境化学、生物学以及物理学等很多学科领域。故 CA 自称是“打开世界化学化工文献的钥匙”。

对于 CA on CD 一般提供四种检索途径：Browse（索引浏览式检索）、Search（词条检索）、Substance（化学物质名称检索）、Formula（分子式检索）。此外还有相关词检索、登记号检索等。

10. European Patent

由欧洲专利局及其成员国提供的免费专利检索数据库，该数据库收录时间跨度大、涉及欧洲以及世界上 70 多个国家的专利文献数据，可检索到著录项目、说明书全文、同族专利、法律状态等内容。

数据库提供了以下四种检索方式：快速检索、高级检索、专利号检索、分类号检索。在主页（http://www.epo.org）的每一种检索方式旁都配有快速帮助信息，能有效地指导用户完成简单检索。值得注意的是，在关键词检索时最多运行四个词进行逻辑组配检索。数据库的专利全文以 PDF 图像格式显示。

虽然上面介绍的数据库所包含的文献资料已不少，但它们大多数是需要支付费用的（校图书馆已购买，可免费使用），如遇到查不到的文献，可能还需要一些辅助工具来帮助检索。如常用的外文搜索引擎谷歌（Google）、雅虎（Yahoo）等，或到有关材料方面的论坛如小木虫学术论坛、材料科学论坛等搜索、求助，可能对你有意想不到的帮助。

UNIT 7　科技英语阅读

Lesson 18　科技术语构成

一、合成法

① 名词+名词　　initiation reaction, point defect, heat capacity
② 形容词+名词　　thermal conductivity, biological degradation, ionic bond
③ 动名词+名词　　lasing action, breaking strength, alternating copolymer
④ 过去分词+名词　　conjugated polymer, absorbed energy, crosslinked reaction
⑤ 名词+动名词　　diffraction grating, materials processing, injection molding
⑥ 名词+过去分词　　sliver-plated, mass-produced, ceramic glazed
⑦ 名词+形容词　　ice-cold, lead-free, water-proof
⑧ 形容词+形容词　　dark-green, light-blue, red-hot

二、派生法

1. **前　缀**

（1）表示否定意义的前缀

ab-　　abnormal, abuse, abend
de-　　degradation, dehydration, decolor
dis-　　discontinuity, disorder, dissimilar
in-(im-,il-,ir-)　　inorganic, impurity,illegal, irreversible
mis-　　misdirect, misunderstand, mistake
non-　　nonmetal, nonconductor, nonageing
un-　　unbending, unstable, uncolored

（2）表示某种意义的前缀

auto- 自己，自　　automation, autocatalysis, autorotation
electro- 电　　electrochemical, electroptic, electropositive
hetero- 异，杂　　heterochain, heteropolar, heterogeneous
homo- 同　　homonuclear, homogeneous, homopolar
hydro- 水或氢　　hydrogenation, hydroxy, hydrolysis

phono- 声，音　　phonograph, phonology, phonological

photo- 光　　photoresist, photochemistry, photoirradiation

（3）表示数量的前缀

bi-　二　　bisphenol A, binary, bicycle

cent-　百　　century, centigrade, centimeter

deca-　十　　decade, decagon, decagram

deci-　十分之一　　decimal, decimeter, deciare

di-　二　　dioxide, diode, dimer

hecto-　百　　hectometer, hectogram, hectowatt

hept-　七　　heptane, heptagon, heptene

hexa-, sex-　六　　hexagon, hexanol, sexangle

kilo-　千　　kilogram, kilowatt, kilometer

mega-　兆,百万　　megaohm, megacycle, megadyne

micro-　微　　micrometer, microscope, microwave

mono-　单　　monoxide, monomer, monolithic

nano-　纳米，毫微　　nanometer, nanoscale, nanotube

nona-　九　　nonanol, nonane, nonagon

oct-　八　　octane, octagon, octahedral

penta-　五　　pentagon, pentane, pentagram

quadri-　四　　quadrivalent, quadricycle, quadrode

tetra-　四　　tetrachloride, tetrahedral, tetragonal

tri-　三　　triangle, trimer, tricar

（4）表示方位的前缀

by-　旁边的，次要的　　byproduct, byeffect, byroad

dia-　通过，横过　　diameter, diagram, diaphanous

ex-　除去，离开　　expose, extract, exhaust

in-　在内，向内　　inlet, inside, input

inter-　之间　　intermediate, interchange, interaction

intro-　向内，向中　　Introduction, introvert, intromit

mid-　中　　midway, midposition, midland

out-　超过，向外　　outbreak, outflow, outside

pro-　向前　　promote, propel, progress

tele-　远　　telephone, telescope, telegram

trans-　横过，转移　　transport, transparent, transplant

under-　在……之下　　underground, underline, undersea

(5) 表示程度的前缀

equi- 同等	equivalent, equilibrium, equitable
fore- 前，先于	forecast, foreword, foreground
hemi- 半	hemisphere, hemicycle, hemipyramid
semi- 半	semiconductor, semicircle, semiautomatic
hypo- 低，次	hypochlorite, hypophosphite, hypotrophy
multi- 多	multiply, multiphase, multiform
over- 超过，过分	overload, overflow, overfulfil
per- 高，过	peroxide, perchloride
poly- 多，聚	polyatomic, polytechnic
post- 在后，补充	post-graduate, post-doctor
pre- 在前，预先	prearrange, prepare
re- 再次，重复	reflect, refract, review, repair, revise
sub- 次于，在下，低	subway, subsonic, subheading
super- 超	supermarket, supersaturate
ultra- 超	ultra-violet, ultra-sonic wave
under- 不足	underexposure, undercharge

(6) 表示反对或逆的前缀

anti-	antifreeze, antirust, antibody
counter-	counterbalance, counterclockwise, counteract
contra-	contrast, contradistinction, contravene

2. 后 缀

(1) 构成名词的后缀

-age	shrinkage, breakage, wastage
-ance	importance, abundance, resistance
-ancy	constancy, occupancy, accountancy
-ant, -ent	reactant, refrigerant, substituent,, solvent
-cide	germicide, pesticide, bactericide
-ence	difference, competence, adherence
-ency	efficiency, deficiency, solvency
-ics	electronics, physics, mechanics
-ism	mechanism, isomerism, magnetism
-ist	physicist, scientist, chemist
-ization	polarization, ionization, polymerization

-ness	hardness, thickness, stiffness
-or, -er	operator, compositor, printer, worker,
-tion, -sion	composition, oxidation, possession, corrosion
-ty	density, stability, ability

（2）构成形容词的后缀

-al	material, crystal, experimental
-ar	molecular, solar, linear
-ble	reversible, soluble, flexible
-ed	colored, advanced, aged
-fold	twofold, thousandfold, manifold
-ful	useful, powerful, truthful
-ic	atomic, alcoholic, dynamic
-less	colorless, waterless, rootless
-ous	amorphous, porous, ferrous
--tive	active, alternative, attractive
-y	earthy, dusty, rainy

（3）构成动词的后缀

-ate	accelerate, congregate, concentrate
-en	brighten, harden, lengthen
-fy	modify, simplify, classify
-ize	modernize, polymerize, delocalize

Lesson 19 科技英语论文阅读

以一般科技英语论文为例来说明阅读步骤。

1. 标 题

阅读一篇科技英语论文，首先看到的是论文的标题，从标题可以粗略地判断出与所需的文献内容是否一致。

2. 作 者

初读者往往容易忽视作者信息，只有列参考文献时才会把它抄下来。其实在阅读论文前应该看一下作者，作者如果有读者自身领域内的权威人士，文章的可读性就大大增强了；如果能够追踪该作者的系列论文，可获得其多年的研究成果、领先程度以及可能的发展方向。这样会对读者有所启迪，并确定自己的研究方向。

3. 论文摘要与关键词

摘要是一名读者取舍一篇论文的关键。只有论文标题，一般是满足不了读者的需要，因此读者在找到感兴趣的论文标题之后，急于阅读的就是摘要。通常在网上或是校园网都可以找到论文的摘要，如果发现该论文与自己的研究相关，就可以进一步查阅全文。故摘要可使读者非常直接地接触文章的核心内容，并很快就可决定这篇论文是否有参考价值，有无阅读完全文的必要。关键词可帮助查阅相关资料。

4. 引　言

引言部分可以放在实验部分和结果与讨论之后再阅读。在进一步的研究工作结束时，引言部分不仅可以提供理论方面的素材，而且还可以提供许多其他的文献资料，这样既可以充实自己在该方向的理论知识，又可以顺利地完成论文撰写。

5. 实验部分

首先，对照论文介绍的实验条件判断是否具备相同或类似的仪器设备，从而确定类似的课题研究是否可以开展，做到知己知彼。其次，还可以从实验过程中看到自己的不足以及改进的方法。此外，还可参照实验部分的清单购置所需要的仪器、设备和材料。

6. 结果与讨论

结果与讨论部分需要反复细致阅读，有时需要阅读多遍，甚至将论文中的结果与自己的实验结果反复对照分析。虽然如此重视某些论文或数据，可他人所得出的这些结果未必准确无误，需要验证。也就是说别人的数据可以拿来参考，但切不可作为衡量自己工作的标准。

表格可以清楚地将数据进行比较，尤其当数据较繁杂时，表格非常有利于查找某一个或某一类数据，而且可以发现在某种特定的实验条件下，实验数据的变化规律、独有的特性，列表可以将很细微的差别反映出来。数据的异常变化（如一系列均匀变化的数据中，有一个或几个数据发生大幅度变化）如果不是实验误差造成的，就要给予特别的注意，因为这些异常很有可能是反映物质本身特性的重要特征。故在发现异常数据之后，要去翻阅讨论部分，了解作者的解释或是独到的见解。

讨论是全篇论文的精彩段落，一篇论文的新颖性、独创性都体现在这一部分。仔细阅读理解作者的思路、洞察力、逻辑结构和基础知识的运用并从中有所启发，还可以从作者那里汲取优点，升华成为自己的方法，指导并运用于实验和自己的科技写作之中。

7. 结　论

结论部分是将论文中的发现和能够形成结论的内容放在一起归纳。实际上，结论的大部分内容已出现在摘要、结果与讨论之中。

8. 致　谢

这部分是作者向提供帮助的个人或单位表示感谢。

9. 参考文献

虽然它只是罗列了一系列他人的论文，然而它在实验方法、实验步骤、实验条件及实验数据的重复等方面，却提供了不可多得的帮助；同时它还可以扩展读者相关领域的知识面。

UNIT 8　科技英语论文写作

Lesson 20　科技英语论文构成

在撰写科技论文之前，作者必须要回答以下 4 个基本问题：你为何要开始（Why did you start）?你做了什么（What did you do）?你发现了什么（What did you find）?它的意义是什么（What does it mean）?这 4 个问题在论文中有固定的格式来阐述和回答，即论文的 IMRAD 结构：引言（Introduction）、材料与方法（Materials and methods）、结果（Results）和（And）讨论（Discussion）；再加上标题（Title）、作者的姓名与工作单位（Authors and their Companies）、摘要（Abstract）、关键词（Key words）、致谢（Acknowledgements）和参考文献（References）就构成了一篇完整的论文。

1. 标　题

标题是论文的画龙点睛之处，它以最精炼的单词和最准确的逻辑组合来表达论文中最深刻的特定内容，反映研究范围和深度，即标题应“以最少数量的单词来充分表述论文的内容”。标题主要功能是让读者了解本文的研究领域和本文区别于其他研究的特征。标题撰写要求清楚明了，标题中每个词的每一个字母应大写（介词和冠词可以除外）或全部字母均大写，同时注意标题不是一个句子，不应有谓语动词和标点符号。

2. 作　者

作者的姓名与工作单位一般有三重意思，一是确定作者在科研工作中所起的作用，通常从排名的顺序可以反映出来；二是文责自负，论文内容若与他人论文相近或是有侵权甚至是法律纠纷，应由作者承担责任。不论从上述哪一种意义上讲，都要求作者将（包括排序）及工作单位写清楚；三是便于读者与作者联系。

3. 摘要与关键词

论文摘要一般由 100 ~ 200 个单词组成，其作用是在标题基础之上对论文进行进一步的介绍。论文摘要是论文主要观点的概括总结，它以最少的文字总结论文的主要观点和内容，是全文的浓缩和精华所在。摘要用词简练，能够简明扼要、准确地概括出全文信息。摘要内容包括论文的研究目的、意义、方法、重要结果和结论。

论文摘要后面紧随 3 ~ 5 个关键词，它是摘要内容的浓缩，用来反映 SCI 论文主题内容最重要的词、词组或短语。关键词可供读者在数据库内检索文献，合理地选择关键词能够增加论文的检出率，提高读者阅读论文的兴趣，增加论文和期刊的被引用率。

4. 引　言

引言部分主要讲述研究工作的目的、范围、相关领域的前人工作和知识空白、理论基础和分析、研究设想、研究方法和实验设计、预期结果和意义等。引言通常用来介绍与本项研究有关的发展历史以及相关理论，是他人研究的概括和本文研究的理论依托。

一般来讲，引言部分可以分三个层次描述。一是简短概括研究领域的背景知识，即对历史问题进行回顾，讲述目前已掌握的知识点和尚存在又未解决的问题；二是总结目前研究现状，陈述本研究的目的和范围；三是重点描述本论文研究目的、研究工作、方法、预期结果及意义。

5. 材料与方法（实验部分）

材料与方法描述了研究所用的材料、方法以及研究实施的基本过程和流程，它不仅回答了实验怎样做的问题，又为下文结果的顺利得出提供物质前提和基础，在论文中提到承上启下的作用。这部分内容不仅可以帮助读者了解实验的进展过程，为他人重复研究提供依据；同时，审稿人还可以依据论文中描写的实验对象、试剂和仪器的合理性和可靠性来判断论文结果是否可信。

6. 结果与讨论

结果与讨论是一篇论文最重要的部分，一篇论文精彩与否大多可以从该部分体现出来。结果与讨论能够充分表现出作者对所研究领域知识背景掌握的情况、理论功底，同时还能看出作者的分析判断能力。从这一节的结构布局还可看出作者的逻辑思维是否做到分析的步步深入，有些资深作者甚至能够做到同一现象从诸多不同的角度来进行分析，使读者享有耳目一新的感觉。

这一部分是全文的精华，运用实验得到的数据，采用文字、图和表展示，结合已有理论，经过周密分析和严密逻辑推理，最终演绎出新的结论或发现。其中结果中的数据不一定是原始数据，可以是经过计算加工后的结果。在前面的工作基础上，根据表格和曲线图描述的变化规律及理论与实际产生差距的原因，做出分析讨论、解释或得出新的发现。

7. 结　论

结论与摘要前后呼应，使整篇论文形成一个整体。结论部分罗列了研究结果和推出的结论。这一节文字少，但很精练，通常只有一个或两个自然段，但要求在这一节体现出研究的科学性和新颖性。

结论中阐述的内容通常有：

（1）作者本人研究的主要认识或论点，其中包括最重要的结果、结果的重要内涵、对结果的说明或认识等；

（2）总结性地阐述研究结果可能的应用前景、研究的局限性及需要进一步深入的研究方向。

8. 致 谢

致谢是对论文完成过程中做出贡献但未列入作者名单的人（如导师、同事、同学、协助准备测试实验的人或单位等）以及项目资金的提供单位等表示感谢。

致谢除了表达道义上的感激外，也是尊重他人贡献的表示。致谢中通常包括的内容有：

（1）感谢任何个人或机构在技术上的帮助，包括提供仪器、设备或相关实验材料，协助实验工作，提供有益的启发、建议、指导、审阅，承担某些辅助性工作，等等；

（2）感谢外部的基金帮助，如资助、协议或奖学金，有时还需标注资助项目号、合同书编号等。

9. 参考文献

参考文献是指为撰写或编辑论文而引用的有关文献信息资源，它不仅表明论文的科学依据和历史背景，而且强调作者在前人研究基础上的提高、发展和创新所在。参考文献中所列是论文作者仔细阅读过并认为非常有参考价值的论文。引文排列以在论文中出现的先后为序，文献序号应与论文中的编号一致。论文中的文献号，当有作者名时放在最后一个作者的右上角，如无作者，则放在该段引文结束的右上角。

Lesson 21 科技英语论文撰写

外刊论文撰写一般有五个基本要求，即 5C：正确（correctness）、清楚（clarity）、简洁（concision）、完整（completion）和一致性（consistency）。

初次写作科技论文起笔时会感觉无从下手，建议初学者按以下顺序着手。

（1）阅读和研究文献，特别是实验室过去发表的论文，了解论文格式、描述方法、领域的前沿和面临的问题等。

（2）先从自己熟悉的部分开始写作，比如先写结果部分，整理实验数据，将自己的实验结果通过列表、画图罗列出来，这部分内容自己熟悉，也是一篇论文的核心。

（3）根据实验结果，针对性写材料与方法。这部分内容需与实验结果相匹配，不能有遗漏也不能有多余的实验，平铺直叙地描述实验所用试剂、纯度、仪器、型号以及对重要实验步骤。

（4）写前言部分，在完成前一部分时阅读了很多材料，再添加一些研究背景以及本研究的重要性、特殊性及构思，就构成了论文的引言部分。

（5）写讨论部分，需要大量阅读理论知识和相近的前人工作、了解他人的成就，总结自己的工作，充分利用现有理论以及自己所掌握的知识对实验结果进行分析讨论，证明结果的正确性和可靠性。

（6）写摘要和结论，一个处于论文的开始，另一个则处于论文的结尾，两段遥相对应，而且内容也几乎是对应的。一段提出将要面临和解决的问题，另一段在经过详细讨论后将已解决的问题总结提炼并形成结论。

（7）写标题，纵观全文将其中心思想用一句简单的话描述出来。标题是论文中最难写的一句话，因为它是所做工作的高度概括和集中体现，既要表达本文的中心思想又要吸引同行读者。

（8）对整篇论文反复修改，论文写作是一个需要下很多功夫的过程，用词、组句、文献和数据的阐述都是需要经过多次反复推敲修改。

（9）仔细阅读期刊的 Guide for Authors，按期刊要求格式整理文献，摘要和其他主体部分。

（10）投稿之前写 cover letter。

一、标　题

标题是正确反映课题研究范围、深度和创新性的最鲜明、最精炼的概括和总结，其构成要素包括研究对象、研究方法、研究目的及主要结果和结论四个方面（四个要素是否同时出现应根据需要和具体情形而定），即通过标题就能使读者和审稿人了解论文的中心内容。一个好标题，既具有简练醒目、引人入胜的特点，又概括了整篇论文的内容，可以吸引读者阅读。

科技论文英文标题撰写 ABC：Accuracy—准确、Brevity—简洁、Clarity—清楚。

标题要准确地反映论文的主要内容。作为论文的"标签"，标题既不能过于空泛和一般化，也不宜过于繁琐。如题名吸引人，读者就有可能进一步阅读摘要或全文，甚至下载并保存，反之则失去读者。题名中准确的"线索"（keys）对于文献检索也至关重要。目前，大多数索引和摘要服务系统都已采取"关键词"系统，因此，题名中的术语应是论文中重要内容的"亮点"（highlight words），并且易被理解和检索。为确保题名的含义准确，应尽量避免使用非定量的、含义不明的词，如"rapid"、"new"等；并力求用词具有专指性。

标题长短应根据论文内容而定，不宜太短、也不宜太长，12 个词左右最好，一般不超过 20 个词，若能用一行文字表达，就尽量不要用两行，必要时可加副标题。虽然要求字数少，但不要使用缩写，因为缩写往往会让读者感到困惑，猜不出标题的内容。

除此之外，标题应清晰地反映论文的具体内容和特色，明确表明研究工作的独到之处，力求简洁有效、重点突出。为表达直接、清楚，以便引起读者的注意，尽可能地将表达核心内容的主题词放在题名开头。为方便二次检索，标题中应避免使用化学式、上下角标、特殊符号（数字符号、希腊字母等）、公式、不常用的专业术语和非英语词汇（包括拉丁语）等。

例 1

Characterization of Monolayers and Langmuir-Blodgett Multilayers of Stearic Acid

这一题目明确地表明单分子和多分子层的特性研究，采用的方法是 L-B 制备薄膜技术，研究对象是硬脂酸。题目明确、内容具体，使读者很容易了解论文介绍的关键内容，很快引起同行的兴趣，产生进一步阅读的欲望。

如果一个短标题不足以概括论文内容或是系列型论文，可以考虑增加副标题。

例 2

Investigation of Electrical Properties of L-B Multilayers of Dioctadecyldimethylammonium chloride (Ⅰ)-Conductivity

作为下一篇论文只需更换（Ⅰ）为（Ⅱ）及连接号后面的内容就可以了。

二、作　者

作者署名标志着作者对文章中涉及的内容所做的贡献，排名次序也意味着所做的贡献大小。对作者来讲是一种荣誉，同时也表明文责自负，对他人来讲是今后文章索引的一个线索。参与署名的作者一般应参与项目的设计和实验工作，对数据处理、理论研究以及论文撰写做出一定的贡献，而且需要阅读过全文并同意发表。

根据现行国家标准（GB/T 16159—1996）规定：汉语人名在英译时采用汉语拼音的写法，姓和名分开来写，且开头字母均大写，双名的拼音一般要写在一起，姓可在名前也可在名后，如 Wang Xiaoming；若名字是三个字还可以用缩写的形式给出，如 Wang X. M.；在多名作者的时候，名字以逗号分开，而最后两作者的名字以 and 相连，如 Wang Xiaoming, Li Dongfang and Zhang Mingming。

在作者下面还要写明作者的工作单位，多名作者同属一个单位就不用重复；而从属于多个单位时就需要分别标明并出各自单位。联系人的名字右上角一般加*号，也称为通信作者，并给出联系地址、电话和 E-mail 地址等。

以下就是作者部分常用形式的一个实例。

Wang Xiaoming[a*], Li Dongfang[b] and Zhang Mingming[a]

[a]School of Materials Science and Engineering, Chongqing Jiaotong University, Chongqing 400074, PR China.

[b]Chongqing Communications Research and Design Institute, Chongqing, 400067, PR China.

[*]Corresponding author. Tel: +86 12345678900. E-mail address: sunny123456 @126.com (Wang X. M .).

三、摘要与关键词

论文摘要（Abstract）是作者所做工作的高度概括，是标题的扩充，其内容可含有研究工作的某些细节或是区别于他人的精彩之作。通常读者在找到感兴趣的标题之后，急于阅读的就是论文摘要，如觉得可取，就会去下载和阅读。因此在查阅相关论文时，摘要可使读者非常直接地接触论文的核心内容，并很快就可以决定这篇论文是否有参考价值，而无须阅读完整篇论文。在参加国际会议时，大会也会根据文章中的摘要决定讨论组的划分以及在论文集中该论文应归属的类别。

摘要的特点是简明扼要，内容一定要简练而又说明问题。学术论文摘要一般在200～300个单词，应包括研究目的、研究对象、研究方法、研究结果、结论及其适用范围这6项内容。

在撰写摘要的过程当中一般需要遵守下列原则：

（1）为确保简洁而充分地表述论文的IMRAD，可适当强调研究中的创新、重要之处（但不要使用评价性语言），尽量包括论文中的主要论点和重要细节（重要的论证或数据）。

（2）使用简短的句子，用词应为潜在的读者所熟悉，表达要准确、简洁、清楚，注意表述的逻辑性，尽量使用指示性的词语来表达论文的不同部分（层次），如使用“We found that...”表示结果，使用“We suggest that...”表示讨论结果的含义等。

（3）应尽量避免引用文献，如若无法回避使用引文，应在引文出现的位置将引文书目信息标注在方括号内，如确有需要（如避免多次重复较长的术语）使用非同行熟知的缩写，应在缩写符号第一次出现时给出全称。

（4）为方便检索系统转录，应尽量避免使用化学结构式、数学表达式、角标和希腊文等特殊符号。

（5）查询拟投稿期刊的作者须知，了解其对摘要字数和形式的要求。

写作过程当中，不但要修改论文的本身，也要不断地调整摘要中的内容。所以，在论文完成之后再写摘要是很正常的，会更具有概括性。

关键词（Key words）是论文信息高度的概括，代表了论文的中心内容和特征。关键词有助于读者进一步了解论文的主要研究方向并加深读者对论文内容的理解。关键词的另一个重要用途是论文检索，关键词输入计算机或是网络系统当中，可使读者利用关键词迅速找到这篇文章。

关键词需从论文中选出，多数来自标题或摘要，一般要根据论文的内容或期刊要求选3～10个。关键词之间用分号隔开，最后一个词不要有标点符号，且每一个实词的第一个字母需大写。注意数学公式、化学式不可以用作关键词。

四、引　言

引言（Introduction）的关键就是一个“引”字，它是说明论文写作的背景、理由、主要研究成果及与前人工作的关系等，目的是引导读者进入论文的主题，并让读者对论文中将要阐述的内容有心理准备。因此，引言有总揽论文全局的重要性，也是论文中最难写的部分之一。

引言部分撰写一般先从大面上起笔，然后再转到本论文的研究工作，阐明现在的问题以及如何去研究这个问题。即按照①某个领域现状分析，②论文研究工作的意义，③存在的问题，④本文进一步研究方法四个步骤完成。在引言写作之前，作者一定要系统查阅文献，对前人的研究基础有全面掌握后再着手撰写，使得引言描述层次清楚，内容全面，做到言简意赅，重点突出。

引言部分不宜过长，2～4个自然段即可，有的论文只有一段。引言字数不等，一般在200～300个单词，要求能将引言的功能全部体现出来。引言撰写时应注意以下基本要求：

（1）尽量准确、清楚且简洁地指出所探讨问题的本质和范围，对研究背景的阐述应繁简适度。

（2）在背景介绍和问题提出中，应引用“最相关”的文献以指引读者，优先选择引用的文献包括相关的经典、重要和最有说服力的文献。

（3）采取适当方式强调作者在本次研究中最重要的发现或贡献，让读者顺着逻辑的演进阅读论文。

（4）解释或定义专门术语或缩写词，以帮助编辑、审稿人和读者阅读与理解。

（5）适当地使用“I”“We”或“Our”，以明确地指示作者本人的工作，如：最好使用“We conducted this study to determine whether...”，而不使用“This study was conducted to determine whether...”。

（6）叙述前人工作的欠缺以强调自己研究的创新时，应慎重且留有余地，可采用类似如下的表达：To the author’s knowledge...，There is little information available in literature about...，Until recently, there is some lack of knowledge about...等。

五、材料与方法（实验部分）

材料与方法部分要向读者证明研究手段的先进性、所得数据的可信性以及实验过程的合理性。因此在介绍实验中所用的主要仪器、设备、材料及药品时，仪器设备要注明型号，最好有厂家的名字（先进性）；材料和药品要注明纯度（如分析纯或化学纯等）、浓度，最好能够写出供货商（可靠性）。

材料与方法部分的写作方法应注意以下几点：

（1）用小标题来组织材料与方法部分，按合理的顺序来讲述实验操作步骤。

（2）时态采用过去时，以第三人称和被动句为主，偶尔使用第一人称。

（3）特别注意实验要定量。对用量、时间、温度等都要具体写明。

（4）明确写出具体条件，试剂、溶液、所用仪器型号、生产厂家等，避免使用含糊的代词。

（5）写明数据处理和分析方法

六、结果与讨论

在经过引言和实验部分的铺垫后，出现的是论文的核心部分——结果与讨论。有些论文将这两部分分开来写，成为两个独立的章节。因此，在撰写科技论文时也有两种不同的方法：第一种是把要讨论的部分分成几个部分，而每一部分都有结论和讨论，彼此之间既相互独立又相互联系，而且后面还可以引用前面部分所分析出的结果。这样的写法思路清晰，简单易懂，但整体性不强，因此在分析或讨论的时候必须加上概述性段落使各段之间有机地结合起来。第二种是先集中介绍实验数据或所得结果，而后对结果进行分析，这样做可以纵观结论，得出整体概念，但缺点是获得的结论往往分析得不够彻底，有时还会遗漏一些细节，使读者产生疑问，同时还使可信度降低。

作者在撰写初稿时最好将二者分开撰写，然后根据需要和编辑的建议来决定是否合并。

1. 结　果

“结果”是作者贡献的集中反映，是整篇论文的立足点，因此也可以说是论文中最重要的部分。论文前部分（引言、实验部分）是为了解释为什么和如何获得这些结果，后部分（讨论）则是为了解释这些结果的蕴含。

结果中通常包括的内容主要有：

（1）结果的介绍：即指出结果在哪些图表中列出；

（2）结果的描述：即描述重要的实验或观察结果；

（3）对结果的评论：即对结果的说明、解释、与模型或他人结果的比较等。

写作要点：

（1）对实验或观察结果的表达要高度概括和提炼；

（2）数据表达可采用文字与图表相结合的形式；

（3）适当说明原始数据，以帮助读者理解；

（4）文字表达应准确、简洁、清楚。

2. 讨　论

“讨论”的重点在于对研究结果的解释和推断，并说明作者的结果是否支持或反对某种观点、是否提出了新的问题或观点等。因此撰写讨论时要避免含蓄，尽量做到直接、明确，以便审稿人和读者了解论文为什么值得引起重视。

讨论通常先简要回顾研究目的和重要结果，然后再讨论具体结果的蕴含及科学意义或贡献。因此，讨论的开始部分范围通常比较窄，然后逐渐变得比较宽泛。

讨论的内容主要有：

（1）回顾研究的主要目的或假设，并探讨所得到的结果是否符合原来的期望，如果没有，需分析原因；

（2）概述最重要的结果，并指出其能否支持先前的假设以及是否与其他学者的结果一致；如果不一致，分析原因；

（3）对结果提出说明、解释或猜测，根据这些结果，能得出何种结论或推论；

（4）指出研究的局限性以及这些局限对研究结果的影响，并建议进一步的研究题目或方向；

（5）指出结果的理论意义（支持或反驳相关领域中现有理论，对现有理论的修正）和实际应用。

写作要点：

（1）对结果的解释要重点突出，简洁、清楚；

（2）推论要符合逻辑，避免实验数据不足以支持的观点和结论；

（3）观点或结论的表述要清楚、准确；

（4）对结果科学意义和实际应用效果的表达要实事求是，适当留有余地。

从论文写作顺序上讲，也可以采用两种方式：一种大体上是按照研究工作进程的时间顺序进行描述。一般研究工作就是按照一种循序渐进的方式向前推进，总是在研究工作当中不断地总结、发现并提出更深层次的研究内容。所以在开始写这一部分时要对其结构和层次进行介绍，然后依次对各个层面进行介绍，这样层次感强、段落清楚而且有从感性认识到理性认识升华的过程。让读者跟着作者的思路走更容易被接受，而且写起来也较为容易下笔。另一种方法是按照逻辑顺序进行排列，这种方法将理论与实践融合在一起并加以提炼，因此认识上是一种由低级向高级演变，而不是记叙研究时间顺序的过程。

七、结　论

结论是全篇论文的概括和总结，撰写结论时要注意以下几个问题：

（1）列入结论当中每一句话，首先要考虑新颖性，如果与他人的结果雷同，则毫无意义；

（2）要突出本文对该领域的贡献并尽量囊括所有在讨论当中所得的结果，切勿遗漏，否则会减轻论文的分量，有时甚至埋没了所做出的贡献；

（3）结论部分要与摘要部分呼应，因此撰写时最好以摘要部分作为参考，如此时摘要部分尚未写好，应根据需要安排好两部分各自的内容，摘要提出的是论文将要讨论的问题和可能的创新点，而结论则是论文创新和贡献的总结；

（4）结论部分不宜过长，通常只有 1～2 个自然段，也可根据需要适当加长。时态多用一般现在时，有时也可用过去时。

总之，撰写结论时不应涉及前文所不曾指出的新事实，也不能在结论中简单地重复摘要、引言、结果或讨论等章节中的句子（表 8-1）；或者叙述其他不重要甚至与本次研究没有密切联系的内容。

表 8-1 “结果”“讨论”与“结论”中应分别侧重的内容

结　果	讨　论	结　论
介绍研究结果(必要时应使用图、照片、表格等形式表述研究发现或实验数据)	探讨所得的结果与研究目的或假设的关系、是否符合原来的期望、与他人研究结果的比较与分析	主要认识或论点
对重要研究结果的描述	对研究结果的解释；如果结果不符合原来的期望，为什么	概述研究成果可能的应用前景及局限性
对重要研究结果的评论（说明、解释、与他人结果的比较等）	重要研究结果的意义(推论)；研究展望	建议需要进一步研究的课题或方向

八、致　谢

写作要点：

（1）致谢内容应尽量具体、恰如其分

致谢的对象应是对论文工作有实质性帮助、贡献的人或机构，因此，致谢中应尽量指出相应对象的具体帮助与贡献。并且，致谢某人可能暗含着某人赞同论文的观点或结论，如果被感谢的人并不同意论文的观点或结论，那么论文公开发表后被感谢的人和作者都会很尴尬。因此，如果是感谢一个思想、建议或解释，就要尽量指明这些内容，以免将被感谢的对象敏感而尴尬地置于为整篇论文承担文责的境地。为表示应有的礼貌和尊重，投稿前应请所有被感谢的对象阅读定稿，以获得允许或默认。

（2）用词要恰当

要注意选用适当的词句来表达感谢之情，避免因疏忽而冒犯本应接受感谢的个人或机构。

（3）致谢形式要遵从拟投稿期刊的习惯和相关规定

参阅拟投稿期刊的“作者须知”和该刊已发表论文的致谢部分，注意其致谢的表达形式和相关要求。尤其是对于感谢有关基金资助的信息，有些期刊要求将其放到“致谢”中，有些则要求将其放在论文首页的脚注中。

国内大部分的英文期刊习惯将基金资助项目的信息作为论文首页的脚注，国外期刊则多将其作为“致谢”的一部分。

以论文首页脚注形式注明基金资助信息的方式举例如下：

Supported by the Major State Basic Research Development Program of China (Grant No. 2001CB309401-05), and the National Natural Science Foundation of China (Grant No. 60171009).

在致谢中注明基金资助信息举例如下：

Acknowledgments This work was supported by the National Natural Science Foundation of China under Grant No. 51508063.

九、参考文献

参考文献列举形式繁杂多样，不同期刊常常要求不同，但对于特定的一种期刊来讲，其参考文献的格式基本上是统一的、规范化的。

如：[1] J. F. Li, K. Wang, B. P. Zhang, L. M. Zhang, *J. Am. Ceram. Soc.* **89**, 706(2006).

其中[1]是文章中引用参考文献的号码，也就是参考文献的顺序。前后的号码一定要一一对应，特别是引用参考文献较多的时候，如硕、博士论文当中。

国外多数期刊的要求为：首先是作者的姓名，要全部写上，不可以用 et al.，接下来是期刊的名称，请注意期刊名称要用斜体字，*J. Am. Ceram. Soc.*是 Journal of the American Ceramic Society 的缩写，**89** 表示该刊的期号，注意是黑体字，后面的 706 表示这篇文章首页所在的页码，括号内的数字表示这篇文章发表的年份。

若引用了一本书，如：[2] R. S. Gerry, “*Physical Chemistry*”, Wiley, Ne York, (1980).

其中 R. S. Gerry 为作者名，“*Physical Chemistry*”是书名，注意要用斜体字同时要用引号，Wiley, Ne York，表示出版商和出版地，括号内的数字表示这本书的出版年份。

国内一些期刊就要求将年份放在期号和页码之前，有的甚至还要求写出参考文献题目的全名。

因此在写作之前应仔细阅读拟投稿期刊的“作者须知”和期刊中的论文，严格按照要求准备参考文献。

UNIT 9 科技英语论文投稿

Lesson 22 科技英语论文投稿

科学界有一句名言：一项科学实验直到其成果发表并被理解才算完成。此名言阐述了两层含义，一是研究成果必须发表，即信息得到传播和交流；二是发表的论文必须被读者接受并理解，即信息被社会认可，产生应有的社会效益和经济效益。

一、选择投稿期刊

在研究工作完成之后，选择一本合适的期刊投稿是必不要少的环节，也是论文发表至关重要的一步。期刊选择合适，论文会很快被接收并发表，节省发表时间；期刊选择不合适，会延误发表时间，甚至不被发表。为了提高投稿命中率，在选择期刊时应综合考虑投稿论文的学术水平和期刊水平匹配度。

1. 客观评价论文水平

客观评价拟投论文的学术水平是投稿的第一步，在进行研究之前或过程中，作者可能已经阅读了大量相关领域的文献资料，在此基础上，对比分析拟投论文与相关领域已发表论文，总结论文体裁类型、创新点，找出不足之处，客观、如实评估论文的科学意义和实用价值，根据目前该研究现状，准确把握拟投论文水平与质量。

2. 选择合适投稿期刊

材料学科英文期刊种类繁多，但每个期刊都有自己的重要研究领域和特色，因此作者必须要十分了解自己研究领域的重要期刊，力求所选择期刊的出版内容与稿件主题确实密切相关。

选择拟投稿期刊时需要综合考虑的因素主要有：

（1）期刊办刊宗旨和范围

办刊宗旨和范围主要指期刊的侧重点和关注所在，作者应认真阅读准备投稿期刊的“作者须知”或“征稿简则”，尤其要注意其中有关刊载论文范围的说明；此外，还应仔细研读最近几期拟投稿期刊的目录和相关论文，以判断稿件主题是否在刊物征稿范围内，稿件内容是否符合刊物要求。

（2）期刊学术影响

期刊影响因子是评价其学术影响最重要的指标，一般来说，影响因子越高，学术水平和影响越高。当然有些专业内部公认的权威期刊，影响因子不一定高，但受同行专家、研究者的认可，因此也不能片面追求影响因子。作者可根据自己的研究领域和学科方向，找出合适的期刊，在对拟投稿论文水平客观评价的基础上，力争在较高水平的期刊上发表。

（3）稿件审阅和发表速度

发表周期是指从编辑部收到稿件到论文发表的时间，它能反映该期刊稿件审阅快慢和发表速度。

不同期刊差别非常大，作者可根据该期刊已发表的论文首页查看稿件的收稿日期（received date）、修回日期（revised date）和论文被接收的日期（accepted date），推断出期刊发表周期的长短。此外，也可以借助有关材料专业的学术科研互动平台，如小木虫论坛中 SCI 期刊点评会提供期刊的基本信息以及发表周期的统计数据，作者可参数统计数据做出相应判断。

（4）稿件录用率

稿件录用率是指编辑部对所投全部稿件的录用比例，一般高水平期刊的录用率较低。因此投稿之前作者可调查拟投期刊的录用率，再结合论文水平层次，做出合理选择。

（5）重视参考文献

稿件中引用的参考文献与稿件内容和主题相关度较高，可重点考察参考文献的发表期刊，初步判定稿件是否适合在这些期刊发表。

（6）期刊发表费用

一般英文期刊不收取审稿费和版面费，但有的期刊需收取版面费如开源 SCI 期刊；另外，在论文中使用彩色图片有些期刊可能需要作者支付附加费和期刊彩色图片印刷费，因此作者可以采用黑白图片代替，并压缩图片大小，减免这部分支出。

二、阅读“作者须知”

几乎所有期刊都有作者须知或投稿指南（Instructions to Authors，Notes to Contributors），有些期刊每一期都刊登简明的“作者须知”，有些则只登在每卷的第一期上，并且不同期刊作者须知的细节可能不尽相同，但目的都是为了给读者提供准备稿件的指南，从而使稿件更容易快捷、正确地发表。

通过“作者须知”可以了解的信息主要有：

（1）刊物的宗旨和范围；

（2）审稿、修回、录用和发表的时限

（3）编辑职责

（4）投稿方式

（5）稿件排版和格式要求

（6）收费情况

（7）其他，如作者署名、利益冲突等

上述内容中大部分是作者在准备稿件时必须要了解的。

三、稿件的录入与排版

认真、细致的录入与排版是稿件准备中必须要履行的工作。大多数期刊编辑部对新收到的稿件首先要进行打印版式方面的审查，最低要求通常有：稿件必须是电脑录排、双倍行距、单面打印，投寄稿件份数、图表设计、体例版式等必须遵守相关期刊的特定要求。如果投稿不能满足上述全部要求，就有可能在编辑部初审后直接退回，或等作者补齐欠缺材料后再送交同行评议。

稿件录入与排版中应注意的方面主要有：

（1）稿件排版格式

稿件的典型排版格式为：标题页、摘要、关键词、前言、材料和方法、结果与讨论、结论、致谢和参考文献、图、图注、表、表的说明、稿件的编排页码等。

稿件每部分都以新的一页开始，论文题目、作者姓名、地址应放第 1 页，摘要、关键词置于第 2 页，引言部分从第 3 页开始，其后每一部分（材料和方法、结论等）都以新的一页开始，插图的文字说明集中放在单独的一页，表和插图（包括图例和文字说明）应集中起来放在稿件的最后，但在正文中要注明相关图表应该出现的位置。

（2）字体与间距要求

稿件一般用 A4 纸（212 mm × 297 mm），Times New Roman 字体、12 号（points）字、单面、通栏、双倍行间距，上、下、左、右的页边距应不少于 25 mm，具体要求详见拟投期刊作者须知。

（3）作者名字和单位书写方式

中文作者姓名有不同的形式，作者姓名严格按照作者须知要求采用规定的表达形式，建议检索曾在拟投期刊上发表过的中国人论文，参照学习。此外，作者有不同身份如第一作者、通信作者、共同第一作者需标注清晰。单位地址按照作者逐人核对，并由小到大的顺序，写出单位所在地和地址。

（4）参考文献核对

作者需核对稿件中引用参考文献的标注与文后的参考文献位置相匹配，同时校核参考文献信息：

作者名、标题、期刊名、年、卷、期和起止页码。

（5）文中缩写标注

文中缩写第一次出现的位置，是否列出全称进行解释，具体可以参考拟投稿期刊近期发表论文的表达形式，后面的行文中只要出现这一名词术语，必须全是缩写形式。

四、稿件的投寄

论文最后一稿按“作者须知”的要求准备好以后，再准备一份投稿信，就可以投稿了。

1. 准备投稿信（covering letter，submission letter）

向英文期刊投稿时，除了正文以外，还需要写一封信，叫投稿信（cover letter），投稿信有助于稿件被送到合适的编辑或可能的评审人手中，一封好的投稿信可以起到很好的作用，就好比求职时的自荐信，要吸引住编辑的眼球，你就成功一半了。

为节省编辑的时间，投稿信要尽量简短明了、重点突出。主要包括以下几方面的内容：

（1）论文题目和所有作者的姓名及单位地址；

（2）论文适宜的栏目以及适合在该刊而不是其他刊物上发表的原因；

（3）论文的主要发现及创新点和重要性；

（4）论文无一稿多投的承诺；

（5）所有刊出作者对论文的确切贡献；

（6）承诺论文内容真实，无伪造；

（7）所有作者均已阅读论文，且同意论文以该版本投稿；

（8）向编辑建议审稿人或因存在竞争关系不宜做审稿人的名单；

（9）通信作者的姓名、详细地址、电话和传真号码、E-mail 地址；

（10）通信作者签名。

2. 互联网在线投稿

在线投稿时，每种期刊的具体要求不一样，因此，在选定期刊后，一定要仔细阅读期刊的投稿须知，根据投稿指南要求，正确完成投稿。互联网在线投稿一般由四个步骤组成：

（1）网上注册。网上投稿前，找到拟投稿期刊的网页后，进入在线投稿链接窗口。第一次向该杂志投稿时，需先进入注册系统，填写基本信息，包括作者姓名、单位地址、联系方式、E-mail 地址等，注册成功后，在邮箱内会收到回执，回执信息提示作者已完成注册，并反馈注册号和密码。注册完成后，进入下一步。

（2）登录投稿系统，上传稿件。通过回执邮件的注册号和密码，登录投稿系统，多数期刊会先要求作者输入稿件的一般信息，包括题目、作者、摘要、关键词，然后上传正文和图表。

（3）确认上传的稿件完整准确。在上传正文和图表前，作者需要通读期刊对上传各部分的具体要求，如正文字数限制、稿件采用的格式、图表的质量、图的格式、分辨率是否够高等，在上传后，需再次确认上传稿件是否符合要求，稿件内容是否完整准确。

（4）完成投稿。按部就班地完成上述操作后，单击投稿发送键（sending 或 submission），投稿即认为已完成，编辑部收到稿件后，作者会收到邮件回执。

五、投稿后的通信（与编辑的联系）

1. 稿件追踪（follow-up correspondence）

稿件投出以后作者最关心两件事，一是编辑是否收到了稿件，二是论文能否被接受。大多数英文期刊收到新稿后，会给作者发一份正式的、收到稿件的通知函。对于不发收稿回执的期刊，作者可以在投稿时附一个署有自己通信地址的明信片，以便编辑收到稿件后通知你。如果投稿 2 周仍无任何有关稿件收到的信息，也可打电话、发 e-mail 或写信给编辑部核实稿件是否收到。

投出的稿件不外乎有 3 种结局：收录(acceptance)、退改(revision)、退稿(rejection)。大多数期刊会尽量在收到稿件的 6 ~ 8 周内形成一个是否录用的决定，如有一些另外的原因要耽搁更长的时间，编辑会给作者一些解释。

2. 稿件退改（revised manuscript）

稿件不做任何修改即被录用通常是很少见的，几乎所有的经审查学术水平达到出版要求的稿件，在发表前都需要退给作者修改其表述及编辑格式，如压缩文章篇幅、重新设计表格、改善插图质量、限制不规则缩写词使用等。然而退给作者修改的稿件并不代表文章已经被接受，文章最终接受与否取决于作者对文章关键性重要内容和表述方式的修改能否达到审稿专家及编辑的要求。

通常退给作者修改的材料包括原稿、审稿专家意见（reviewers' comments）和一封编辑的信（covering letter）。

当作者收到退改稿后，首先应该仔细地阅读退修信（modify letter）和审稿专家意见。然后应考虑能否或愿意接受审稿专家或编辑的意见，修改稿件。

如果退修意见较少，且为非实质性问题，那么应该遵照退修意见认真修改。

如果编辑要求作者对文章作重大修改，那么应记住并非所有审稿专家的意见都是正确的，都必须无条件接受。这时应注意区分以下几种情况：

（1）审稿意见正确，文章中存在重大错误，根据退修意见重写。

（2）部分审稿意见不正确，根据可接受的建议修改稿件；同时，附一封说明信，根据审稿意见进行逐条陈述。如果你的陈述正确且具有说服力，编辑有可能采取妥协态度。

（3）审稿意见几乎完全错误，一位或两位审稿专家和编辑未能读懂或未能很好地理解作者的原意，这时作者可选择两种方法：一是另投他刊，希望自己的论文得到更公正的评价；二是不放弃原投稿期刊，运用自己所掌握的材料或论据，对审稿人的意见进行逐项详尽地申辩（一定不要使用带有敌对情绪的词语），以期望稿件能送给其他审稿人进行再次评审。

作者一定要在编辑规定的时间内将修改稿返回，否则稿件将从被考虑发表的文章中剔除，按退稿处理。

3. 稿件退稿

如果作者收到的是一封退稿信，应仔细阅读退稿信并决定采取何种处理措施。

（1）完全性退稿，在这种情况下再次投稿给同一家刊物或进行申辩都是毫无意义的。如稿件中的确存在严重问题，最好不要改投其他刊物，以免影响作者本人名誉；如稿件中还有值得保留的内容，可将其改写成一篇全新的文章，然后再尝试重新投稿。

（2）稿件包含一些有用的信息，但有些资料有误。首先要仔细阅读稿件和审稿意见，以确认数据是否有严重的错误，若的确是很大的缺陷，应认真弥补这个缺陷，然后再投稿给这家刊物。如果认为是审稿有误，那么，除非能对编辑进行有说服力的证明，否则最好不要将稿件再投给同一家刊物，考虑另投其他类似的刊物。

（3）除了所做的实验有一些缺陷外，稿件基本上是可以被接受的。作者可以按照审稿人的意见进行必要的修正，然后再次投稿给这家刊物，还是有可能被接受的。

六、核改校样

校样（proof）指论文在期刊上发表前供校对用的印刷样。校核的目的是排除各种错误，以使发表出来的论文尽量完美无瑕。因此作者应逐字逐句仔细核校，力争将错误降到最低限度。

1. 核校内容

期刊编辑部发作者校样的目的只是为了让作者纠正校样中可能存在的错误，主要是印刷错误，而不是让作者重写或大修文章。因为在校样阶段改动太大，一是会延误期刊的按时出版；二是有可能因版面调换而出现新的、更大的错误；三是费用较高。因此校样应尽量少改动（仅做必要的改动），在给编辑的信（covering letter）中回答编辑提出的各种问题。另外，校样应在规定的时间内按要求尽快返回编辑部，以免拖延期刊按时出版，或因编辑部等不及校样而使错误不能得到更正。

2. 正确使用校对符号

国外期刊往往要求作者用标准的校对符号（proofreader’s marks）校稿（marking proof），而英美国家使用的校对符号与我国编辑出版界使用的校对符号不完全一致；另外，他们往往使用双重校对系统（double marking system），即不仅在文中需修改的部位做出标记，还在文旁再做标记以引起注意。因此，作者应了解并会使用这些校对符号。

作者除了处理好以上投稿后若干问题外，还应保存好与发表论文有关的一切材料，因为有些期刊在研究论文发表后 5 年内可能要求作者提供原始资料。

最后特别提醒大家：切勿一稿多投！

APPENDIXES

Append. A 元素名称

元素序号	符号	英文名称	中文名称	元素序号	符号	英文名称	中文名称
1	H	Hydrogen	氢	26	Fe	Iron	铁
2	He	Helium	氦	27	Co	Cobalt	钴
3	Li	Lithium	锂	28	Ni	Nickel	镍
4	Be	Beryllium	铍	29	Cu	Copper	铜
5	B	Boron	硼	30	Zn	Zinc	锌
6	C	Carbon	碳	31	Ga	Gallium	镓
7	N	Nitrogen	氮	32	Ge	Germanium	锗
8	O	Oxygen	氧	33	As	Arsenic	砷
9	F	Fluorine	氟	34	Se	Selenium	硒
10	Ne	Neon	氖	35	Br	Bromine	溴
11	Na	Sodium	钠	36	Kr	Krypton	氪
12	Mg	Magnesium	镁	37	Rb	Rubidium	铷
13	Al	Aluminium	铝	38	Sr	Strontium	锶
14	Si	Silicon	硅	39	Y	Yttrium	钇
15	P	Phosphorus	磷	40	Zr	Zirconium	锆
16	S	Sulphur	硫	41	Nb	Niobium	铌
17	Cl	Chlorine	氯	42	Mo	Molybdenum	钼
18	Ar	Argon	氩	43	Tc	Technetium	锝
19	K	Potassium	钾	44	Ru	Ruthenium	钌
20	Ca	Calcium	钙	45	Rh	Rhodium	铑
21	Sc	Scandium	钪	46	Pd	Palladium	钯
22	Ti	Titanium	钛	47	Ag	Silver	银
23	V	Vanadium	钒	48	Cd	Cadmium	镉
24	Cr	Chromium	铬	49	In	Indium	铟
25	Mn	Manganese	锰	50	Sn	Tin	锡

元素序号	符号	英文名称	中文名称	元素序号	符号	英文名称	中文名称
51	Sb	Antimony	锑	85	At	Astatine	砹
52	Te	Tellurium	碲	86	Rn	Radon	氡
53	I	Iodine	碘	87	Fr	Francium	钫
54	Xe	Xenon	氙	88	Ra	Radium	镭
55	Cs	Cesium	铯	89	Ac	Actinium	锕
56	Ba	Barium	钡	90	Th	Thorium	钍
57	La	Lanthanum	镧	91	Pa	Protactinium	镤
58	Ce	Cerium	铈	92	U	Uranium	铀
59	Pr	Praseodymiu m	镨	93	Np	Neptunium	镎
60	Nd	Neodymium	钕	94	Pu	Plutonium	钚
61	Pm	Promethium	钷	95	Am	Americium	镅
62	Sm	Samarium	钐	96	Cm	Curium	锔
63	Eu	Europium	铕	97	Bk	Berkelium	锫
64	Gd	Gadolinium	钆	98	Cf	Californium	锎
65	Tb	Terbium	铽	99	Es	Einsteinium	锿
66	Dy	Dysprosium	镝	100	Fm	Fermium	镄
67	Ho	Holmium	钬	101	Md	Mendelevium	钔
68	Er	Erbium	铒	102	No	Nobelium	锘
69	Tm	Thulium	铥	103	Lr	Lawrencium	铹
70	Yb	Ytterbium	镱	104	Rf	Rutherfordium	鑪/𬬻
71	Lu	Lutetium	镥	105	Ha	Dubnium	𬭊
72	Hf	Hafnium	铪	106	Sg	Seaborgium	𬭳
73	Ta	Tantalum	钽	107	Bh	Bohrium	𬭛
74	W	Tungsten	钨	108	Hs	Hassium	𬭶
75	Re	Rhenium	铼	109	Mt	Meitnerium	鿏/鿏
76	Os	Osmium	锇	110	Ds	Darmstadtium	鐽/𫟼
77	Ir	Iridium	铱	111	Rg	Roentgenium	錀/𬬭
78	Pt	Platinum	铂	112	Cn	Copernicium	鎶/鿔
79	Au	Gold	金	113	Nh	Nihonium	鉨/鿭
80	Hg	Mercury	汞	114	Fl	Flerovium	鈇/𫓧
81	Tl	Thallium	铊	115	Mc	Moscovium	鏌/镆
82	Pb	Lead	铅	116	Lv	Livermorium	鉝/𫟷
83	Bi	Bismuth	铋	117	Ts	Tennessine	鿬
84	Po	Polonium	钋	118	Og	Oganesson	鿫

Append. B 常用英文数目词头

数目	前缀	数目	前缀	数目	前缀	数目	前缀
1/2	hemi, semi	10	deca	20	eicosa	30	triaconta
1	mono, uni	11	undeca, hendeca	21	heneicosa	31	hentriconta
2	di, bi	12	dodeca	22	docosa	40	tetraconta
3	tri, ter	13	trideca	23	tricosa	50	pentaconta
4	tetra, quadri	14	tetradeca	24	tetracosa	60	hexaconta
5	penta	15	pentadeca	25	pentacosa	70	heptaconta
6	hexa, sexi	16	hexadeca	26	hexacosa	80	octaconta
7	hepta, septi	17	heptadeca	27	heptacosa	90	nonaconta
8	octa	18	octadeca	28	octacosa	100	hecta
9	nona	19	nonadeca	29	nonacosa		

举例 153 烷：tri(3)+pentaconta(50)+hect(100)+ane(烷)=tripentacontahectane.

Append. C 氧化物、氢氧化物和过氧化物等

类别	后缀	实例		缩写
氧化物	-oxide ['ɒksaɪd]	氧化钠	sodium oxide ['səʊdi:əm 'ɔksaid]	
	-dioxide [daɪ'ɒksaɪd]	二氧化碳	carbon dioxide ['kɑ:bən daɪ'ɔksaɪd]	
氢氧化物	-hydroxide [haɪ'drɒksaɪd]	氢氧化钙	calcium hydroxide ['kælsiəm hai'drɔksaid]	
过氧化物	-peroxide [pə'rɒksaɪd]	过氧化苯甲酰	benzoyl peroxide ['benzəuil pə'rɔkˌsaɪd]	BPO
		月桂酰过氧化物	Lauroyl peroxide ['lɔ:rəuil pə'rɔkˌsaɪd]	LPO
		过氧化二异丙苯	dicumyl peroxide	DCP
		异丙苯过氧化氢	cumene hydroperoxide ['kju:minˌhaidrəpə'rɔksaid]	
过硫酸盐	-persulfate [pəsʌl'feɪt]	过硫酸钾	potassium persulfate [pə'tæsiəm pəsʌl'feɪt]	
偶氮化物		偶氮二异丁腈	azobisisobutyronitrile [əzɒbaɪsɪsəʊ'bʌtɪrəni:traɪ]	AIBN

Append. D 无机酸、碱和盐

类别	词尾	实例	
不含氧酸	-ic acid [-ik'æsid]	氢氟酸	hydrofluoric acid ['haidrəflu(:)'ɔrik 'æsid]
		氢氯酸	hydrochloric acid [ˌhaɪdrəˌklɒrɪk 'æsɪd]
		氢溴酸	hydrobromic acid ['haidrəu'brəumik 'æsid]
		氢碘酸	hydriodic acid [ˌhaidri'ɔdik 'æsid]
		氢氰酸	hydrocyanic acid [ˌhaɪdrəʊsaɪ'ænɪk 'æsid]
		氢硫酸	hydrosulphuric acid [ˌhaidrəusʌl'fjuərik 'æsid]
不含氧酸的盐	-ide [-aid]	氟化物	fluoride ['flɔ:raɪd]
		氯化物	chloride ['klɔ:raɪd]
		溴化物	bromide ['brəʊmaɪd]
		碘化物	iodide ['aɪədaɪd]
		氰化物	cyanide ['saɪənaɪd]
		硫化物	sulfide ['sʌlfaɪd]
含氧酸	-ic acid [-ik'æsid]	硝酸	nitric acid [ˌnaɪtrɪk 'æsɪd]
		硫酸	sulphuric acid [sʌlˌfjʊərɪk 'æsɪd]
		碳酸	carbonic acid [kɑ:ˌbɒnɪk 'æsɪd]
		硼酸	boric acid [ˌbɔ:rɪk 'æsɪd]
		磷酸	phosphoric acid [fɒsˌfɒrɪk 'æsɪd]
		硅酸	silicic acid [si'lisik 'æsid]
含氧酸的正盐	-ate [-eit]/[-it]	碳酸盐	carbonate ['kɑ:bəneɪt]
		硝酸盐	nitrate ['naɪtreɪt]
		硫酸盐	sulfate ['sʌlˌfeɪt]
		硼酸盐	borate ['bɔ:ˌreɪt]
		磷酸盐	phosphate ['fɒsfeɪt]
		硅酸盐	silicate ['sɪlɪkeɪt]
含氧酸的亚盐	-ite [-ait]	亚硝酸盐	nitrite ['naɪtraɪt]
		亚硫酸盐	sulfite ['sʌlfaɪt]
		亚磷酸盐	phosphite ['fɒsfaɪt]

Append. E　烷烃、烯烃、炔烃和芳烃

类别	词尾	实例	
烷烃 alkane ['ælkeɪn]	-ane　[-ein]	甲烷	methane ['mi:θeɪn]
		乙烷	ethane ['i:θeɪn]
		丙烷	propane ['prəʊpeɪn]
		丁烷	butane ['bju:teɪn]
		戊烷	pentane ['penteɪn]
		己烷	hexane [hek'seɪn]
		庚烷	heptane ['hepteɪn]
		22 烷	docosane ['dɒkəseɪn]
		132 烷	dotriacontadecatane
烯烃	-ene　[-i:n]	乙烯	ethylene ['eθɪli:n]
		丙烯	propylene ['prəʊpəli:n]
		丁烯	butylene ['beti:li:n]
二烯烃 alkadiene['ælkədɪən]	-diene [-daii:n]	丁二烯	butadiene [ˌbju:tə'dɑɪi:n]
		戊二烯	pentadiene [pentə'daɪi:n]
		戊二烯-[1,3]	pentadiene-[1,3] [pentə'daɪi:n]
		戊二烯-[2,3]	pentadiene-[2,3] [pentə'daɪi:n]
		己二烯-[2,4]	hexadiene-[2,4] ['heksədɪən]
炔烃 alkyne['ælkɑɪn]	-yne　[-ain]	乙炔	ethyne ['i:θaɪn] acetylene [ə'setəli:n]
		丙炔	propyne ['prəʊpaɪn]
芳烃 arene [ə'ri:n]		苯	benzene ['benzi:n]
		甲苯	methylbenzene [meθɪl'benzi:n] toluene ['tɒljʊi:n]
		二甲苯 （邻，间，对）	dimethylbenzene [daɪmeθɪl'benzi:n](*o*-,*m*-,*p*-) xylene ['zaɪli:n](*o*-,*m*-,*p*-)
		苯乙烯	styrene ['staɪri:n]

Append. F 醇、酚、醛、酮和醚

类别	词尾	实例	
醇	-ol [-ɒl] -alcohol [-'ælkəhɒl]	甲醇	methanol ['meθənɒl] methyl alcohol ['meθɪ l'ælkəhɒl]
		乙醇	ethanol ['eθənɒl] ethyl alcohol [eθɪ'l 'ælkəhɒl]
		正丙醇	n-propyl alcohol [en 'prəupil 'ælkə͵hɔl] propanol-[1] [p'rɒpənɒl]
		烯丙醇	allyl alcohol ['ælil 'ælkə͵hɔl]
二元醇	-diol [-daioul] -glycol [-'glaɪkɒl]	乙二醇	glycol ['glaɪkɒl] ethanediol [iːθeɪniː'daɪəl] ethylene glycol ['eθə͵liːn 'glaikɔl]
		丙二醇-[1,2]	*α*-propylene glycol [͵ælfə'prəupiliːn 'glaikɔl]
		丙二醇-[1,3]	*β*-propylene glycol [͵beitə'prəupiliːn 'glaikɔl]
杂原子醇	-hydrin [-haidrin]	氯醇	chlorohydrin [klɔːrə'haɪdrɪn]
		溴醇	bromohydrin [b'rəʊmə'haɪdrɪn]
		碘醇	iodohydrin [aɪədəʊ'haɪdrɪn]
		氰醇	cyanohydrin [saɪənəʊ'haɪdrɪn]
硫醇	-thiol [-'θaɪoul]	十二碳硫醇	dodecanethiol ['dəʊdɪkeɪn'θaɪoul]
	mercaptan [mə'kæptæn]	十八碳硫醇	octadecyl mercaptan [ɒk'tædəsɪl mə'kæptæn]
酚	-ol [-ɒl]	苯酚	phenol ['fiːnɒl]
		苯甲酚 （邻，间，对）	crezol(*o*-,*m*-,*p*-) ['kriːzɒl]
醛	-aldehyde ['ældɪhaɪd] -al[əl]	甲醛	formaldehyde [fɔː'mældɪhaɪd]
		乙醛	ethanal ['eθənæl] acetaldehyde [͵æsɪ'tældəhaɪd]
酮	-one [əʊn] -ketone ['kiːtəʊn]	丙酮	acetone ['æsɪtəʊn]
		丁酮-[2]	butanone-[2] [bjuːtɑː'nɒn] methyl ethyl ketone ['meθil 'eθil 'kiːtəun]
醚	-ether ['iːθə]	（二）乙醚	ethyl ether ['eθil 'iːθə] diethyl ether [dai'eθil 'iːθə]
		甲乙醚	ethyl methyl ether ['eθil 'meθil 'iːθə]

Append. G 羧酸、酯、胺、酰胺和砜

类别	词尾	实例	
羧酸	-ic acid [-ik'æsɪd]	甲酸（蚁酸）	formic acid [ˌfɔːmɪk 'æsɪd]
		乙酸（醋酸）	acetic acid [əˌsiːtɪk 'æsɪd]
		丙烯酸	acrylic acid [ə'krɪlɪk 'æsid]
		乙二酸（草酸）	oxalic acid [ɔk'sælik 'æsid]
		丙二酸（缩苹果酸）	malonic acid [mə'ləʊnɪk, 'æsid]
		丁二酸（琥珀酸）	succinic acid [sək'sɪnɪk 'æsid]
		顺丁烯二酸（马来酸）	maleic acid ['meɪlɪk 'æsid] *cis*-butenedioic acid
		反丁烯二酸（富马酸）	fumaric acid [fju'mærik 'æsid] *trans*-butenedioic acid
		十二酸（月桂酸）	lauric acid ['lɔːrɪk'æsid] dodecanic acid
		十八酸（硬脂酸）	stearic acid [sti'ærik 'æsid] octadecanoic acid
		磺酸	sulfonic acid [sʌl'fɔnik 'æsid]
酸酐	-ic anhydride [-ik æn'haɪˌdraɪd]	醋酸酐	acetate ['æsɪteɪt]
		顺丁烯二酸酐（马来酸酐）	maleic anhydride ['meɪlɪk æn'haɪˌdraɪd]
		反丁烯二酸酐（富马酸酐）	fumaric anhydride [fju'mærik æn'haɪˌdraɪd]
酯	-ate [-eit],[-it] -ester [-'estə]	醋酸酯	acetate ['æsɪteɪt]
		醋酸乙酯	ethyl acetate ['eθil 'æsiˌteit]
		醋酸乙烯酯	vinyl acetate ['vaɪnəl 'æsiˌteit]
		丙烯酸酯	acrylate ['ækrɪleɪt]
		甲基丙烯酸酯	methacrylate [me'θækrəleɪt]
内酯	-lactone [-'læktəʊn]	己内酯	caprolactone [kæprəʊ'læktəʊn]
胺	-amine [-ə'miːn]	甲胺	Methylamine [ˌmeθɪlə'miːn]
二元胺	-diamine ['daɪəmiːn]	乙二胺	ethylenediamine ['eθəliːn'daɪəmiːn]
酰胺	-amide [-'æmaɪd]	乙酰胺	acetamide [æsɪ'tæmaɪd]
酰亚胺	Imide ['ɪmaɪd]	酰氯代亚胺	imide chloride ['imaid 'klɔːraid]
内酰胺	-lactam [-'læktæm]	己内酰胺	caprolactam [kæprəʊ'læktəm]
砜	sulfone ['sʌlfəʊn]	乙基、甲基砜	sulfone ethyl methyl
亚砜	Sulfoxide [sʌl'fɒksaɪd]	烯丙基、丁基亚砜	sulfoxide allyl butyl

Append. H 常用有机基团

类别	词尾	实例		
一价基	-yl [-il]	CH_3—	甲基	
		C_2H_5—	乙基	
		C_3H_7—	丙基	
		C_4H_9—	丁基	
		$CH{=}CH_2$—	乙烯基	
		$CH{=}CH{-}CH_2$—	烯丙基	
		$CH_3{-}CH{=}CH$—	丙烯基	
		C_6H_5—	苯基	
		OH—	羟基	
	其他	—COOH	羧基	
		—CN	氰基	
		$-NO_2$	硝基	
		C_6H_5COO—	苯甲酰基	
二价基	-ylene [-illi:n]	$-CH_2-$	甲撑（甲叉，亚甲基）	
		$-CH_2CH_2-$	乙撑（乙烯，次乙基）	
		$-C_6H_4-$	苯撑	
	-ylidene	$C_2H_4<$	乙叉	
	其他	—S—	硫基（硫代）	thio rubber ['θaɪəʊ'rʌbə(r)]

Append. I 常用聚合物

类别	中文名称	英文名称	缩写
乙烯类	聚乙烯	polyethylene[ˌpɒli'eθəli:n]	PE
	聚丙烯	polypropylene [ˌpɒli'prəupəli:n]	PP
	聚氯乙烯	polyvinyl chloride [ˌpɔli'vainil 'klɔ:raid]	PVC
	聚苯乙烯	polystyrene [ˌpɒli'staɪri:n]	PS
	聚醋酸乙烯酯	polyvinyl acetate [ˌpɔli'vainil 'æsiˌteit]	PVAc
	聚乙烯醇	polyvinyl alcohol [ˌpɔli'vainil 'ælkəˌhɔl]	PVA
	聚丙烯酸酯	polyacrylate [pɒljəkrɪ'leɪt]	
	聚甲基丙烯酸酯	polymethacrylate [pɒlɪme'θækrəleɪt]	PMA
	聚甲基丙烯酸甲酯	poly(methyl methacrylate)	PMMA
	聚丙烯腈	polyacrylonitrile [ˌpɒlɪ'ækrələʊ'naɪtrɪl]	PAN
	聚异丁烯	polyisobutene [pɒli:ɪsəʊbju:'ti:n]	PIB
	聚四氟乙烯	polytetrafluoretyhylene	PTFE
双烯类	聚丁二烯	polybutadiene [pɒlɪbju:tə'daɪi:n]	PB
	聚异戊二烯	polyisoprene [pɒlɪ'aɪsəʊpri:n]	PIP
	聚氯丁二烯	polychlorobutadiene[pɒli:klərəʊbju:'teɪdɪən] polychloroprene [pɒlɪ'klɒ(:)rəpri:n]	PCB PCP
烯烃共聚物	乙烯-丙烯-二烯烃三元聚合物（三元乙丙橡胶）	ethylene-propylene-diene monomer	EPDM
聚酯	聚对苯二甲酸乙二醇酯	polyethylene glycol terephthalate	PET
聚碳酸酯	聚碳酸酯	polycarbonate [ˌpɒli'kɑ:bənət]	PC
聚砜	聚砜	polysulfone [pɒli:sʌlf'wʌn]	PSF
聚酰胺	聚酰胺	polyamide [pɒlɪ'æmaɪd]	PA
聚酰亚胺	聚酰亚胺	polyimide [pɒlɪ'ɪmaɪd]	PI
聚脲	聚脲	polyurea [pɒli:u:'rɪə]	PU
聚氨酯	聚氨基甲酸酯	polyurethane [ˌpɒli'jʊərəθeɪn]	PUR PU
聚醚	聚甲醛	polyformaldehyde [pɒlɪfɔ:'mældɪhaɪd]	POM
	环氧树脂	epoxy resin [e'pɔksi 'rezɪn]	EP
其他	酚醛树脂	phenol formaldehyde resin ['fi:nəl fɔ:'mældəˌhaɪd 'rezɪn]	PF
	脲醛树脂	urea-formaldehyde resin [jʊ'ri:əfɔ:'mældəˌhaɪd]	UF
	三聚氰胺树脂	melamine resin ['meləmi(:)n 'rezɪn]	
	不饱和树脂	unsaturated polyester ['ʌn'sætʃəreɪtɪd ˌpɒli'estə(r)]	UP

Append. J　常用溶剂

中文名称	英文名称	缩写
苯	benzene ['benzi:n]	BNZ
甲苯	toluene ['tɒljʊi:n] methylbenzene [meθɪl'benzi:n]	TOL
二甲苯	xylene ['zaɪli:n] dimethylbenzene [daɪmeθɪl'benzi:n]	
苯乙烯	styrene ['staɪri:n]	St
二甲基甲酰胺	dimethylformamide / [daɪme θ ɪl'fɔ:mæmaɪd]	DMF
环己烷	cyclohexane / [saɪkləʊ'hekseɪn]	CYH
四氢呋喃	tetrahydrofuran / [tetrəhaɪdrə'fjʊərən]	THF
氯仿	chloroform / ['klɒrəfɔ:m]	
丙酮	acetone / ['æsɪtəʊn]	ACT
甲醇	methanol['meθənɒl]	Mt
乙醇	ethanol　['eθənɒl]	Et
丙三醇（甘油）	propanetriol [prə'pænətrɪəʊl]	
间甲酚	*m*-cresol ['emkr'esɒl]	
苯酚	phenol ['fi:nɒl]	
甲醛	formaldehyde[fɔ:'mældɪhaɪd]	
乙醚	ether ['i:θə(r)] ethyl ether ['eθil 'i:θə] diethyl ether[dai'eθil 'i:θə]	
二氧六环	dioxane [daɪ'ɒkseɪn]	DOX
吡啶	pyridine ['pɪrɪdɪ(:)n]	
正辛烷	*n*-octane ['en'ɒkteɪn]	
四氯化碳	carbon tetrachloride ['kɑ:bən ˌtetrə'klɔ:raid]	
乙酸乙酯	ethyl acetate ['eθil 'æsiˌteit]	
乙二醇-[1,2]	ethanediol-[1,2] [i: θ eɪni:'daɪəl]	
异丙醇	isopropyl alcohol[ˌaisəu'prəupil 'ælkəˌhɔl] isopropanol [ˌaɪsə'prəʊpənəʊl]	

参考文献

[1] 匡少平，张永恒，李旭东，等. 材料科学与工程专业英语[M]. 北京：化学工业出版社，2003.

[2] 刘科高，田清波. 材料科学与工程专业英语精读[M]. 北京：冶金工业出版社，2012.

[3] 杜永娟. 无机非金属材料专业英语[M]. 北京：化学工业出版社，2002.

[4] 程为庄，顾国芳. 大学专业英语阅读教程（高分子材料）[M]. 上海：同济大学出版社，1999.

[5] 水中和. Introduction to Materials（材料概论）[M]. 武汉：武汉理工大学出版社，2005.

[6] 董亚明，佟方，庄思永，等. 理工科专业英语[M]. 上海：华东理工大学出版社，2003.

[7] 李达，李玉成，李春艳，等. SCI 论文写作解析[M]. 北京：清华大学出版社，2012.

[8] 张俊东，杨亲正，国防，等. SCI 论文写作和发表：You Can Do It[M]. 北京：化学工业出版社，2015.